CULTURE, KINSHIP AND GENES

Also by Angus Clarke

GENETIC COUNSELLING: Practice and Principles (*editor*)

Culture, Kinship and Genes

Towards Cross-Cultural Genetics

Edited by

Angus Clarke
Reader in Clinical Genetics
Department of Medical Genetics
University of Wales College of Medicine
Cardiff

and

Evelyn Parsons
Senior Lecturer
Department of Nursing Studies
University of Wales College of Medicine
Cardiff

palgrave
macmillan

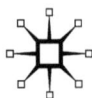

Published by PALGRAVE MACMILLAN
Houndmills, Basingstoke, Hampshire RG21 6XS and
175 Fifth Avenue, New York, N. Y. 10010
Companies and representatives throughout the world

PALGRAVE MACMILLAN is the global academic imprint of the Palgrave
Macmillan division of St. Martin's Press, LLC and of Palgrave Macmillan Ltd.
Macmillan® is a registered trademark in the United States, United Kingdom
and other countries. Palgrave is a registered trademark in the European
Union and other countries.

Outside North America
ISBN 0–333–64702–5

In North America
ISBN 0–312–17499–3

This book is printed on paper suitable for recycling and
made from fully managed and sustained forest sources.

A catalogue record for this book is available from the British Library.

Library of Congress Catalog Card Number: 97–7113

Printed and bound in Great Britain by
Antony Rowe Ltd, Chippenham and Eastbourne

Contents

v

Preface

This volume has arisen out of a meeting held in Abergavenny, Gwent, Wales in March 1994. This meeting, entitled 'Culture, Kinship and Genes', brought together some 60 social scientists and health professionals interested in the impact of genetic issues on members of minority ethnic, religious and cultural groups. The spoken papers delivered at that meeting, and a few written papers that were circulated for discussion, have been developed into the essays published here.

The object of the meeting was to foster interdisciplinary discussion between social scientists and health professionals interested in genetics services. While some psychosocial studies of genetic screening and genetic counselling have been carried out in Britain, these have generally been questionnaire-based psychological studies of the responses of individuals to genetic testing, and they have not focused upon the impact of genetics services on individuals from cultural and ethnic minority groups.

The organisers of the meeting hoped that social scientists interested in other cultural and ethnic groups, such as sociologists and social anthropologists, would be stimulated to embark upon collaborative research studies with health professionals. These studies could examine the knowledge and beliefs about inheritance and inherited disease in different communities, or the responses of individuals and families from different communities to the provision of clinical genetics services.

The potential benefits from such collaborative studies would be two-fold. First, the health service providers of clinical genetics services would have a richer description and therefore a deeper understanding of genetic counselling. This would help them to find ways of improving their services to minority groups. Furthermore, improvements in the genetics services to clients of all ethnic and cultural groups might result from an improved sensitivity to clients' needs. Second, the social scientists would gain valuable insights into the understanding of inheritance and of disease in the social groups studied. They would also witness many different styles of coping with serious disease and handicap, and different attributions of blame and responsibility for misfortune. This area is of such importance to the participating families, and it offers insights into so many areas of social and moral consequence, that studies of cross-cultural genetic counselling have a great potential for deepening our understanding of 'cultural difference' more generally.

Acknowledgements

The editors would like to thank many people for their support; this book would never have appeared without their help. First, we must thank the other enthusiasts who planned the 'Culture, Kinship and Genes' meeting in 1994; the steering group consisted of Elizabeth Anionwu, Rohan Burke, Aamra Darr, Jeanette Edwards, Gulshen Karbani, Theresa Marteau, Bernadette Modell and Nadeem Qureshi, in addition to ourselves. Then we would like to thank Mrs Jean Dunscombe, in the Department of Medical Genetics in Cardiff, who supported us unfailingly in so many ways, and all the participants at the meeting in Abergavenny – those who presented papers and those who contributed in discussion. We would particularly like to thank those participants who came from farthest away – Professors Olu Akinyanju and Jennifer Kromberg. We would like to thank the staff of The Hill Residential College in Abergavenny, who gave us such a warm welcome; Annabelle Buckley and her colleagues at Macmillan for their patience and their trust; and, finally, the organisations who gave us generous financial support – the Wellcome Trust and the Medical Research Council – without whose contributions the meeting might never have taken place, and hence this volume might never have been produced.

Notes on the Contributors

Olu Akinyanju is Professor of Medicine in Lagos, Nigeria. He is also founder and Chair of the Sickle Cell Foundation of Nigeria.

Angus Clarke is Reader in Clinical Genetics at the University of Wales College of Medicine in Cardiff and Honorary Consultant Clinical Geneticist.

Aamra Darr is a medical sociologist with an interest in community genetics and the health care needs of ethnic minorities in Britain. She is the Information and Awareness Officer at the Genetic Interest Group and an honorary research fellow at the Department of Primary Care and Population Sciences of the University College London Medical School.

Charlie Davison is Lecturer in the Department of Sociology, University of Essex.

Merry France-Dawson is a freelance socio-medical researcher based in Oxford.

Susie Godsil is an independent psychoanalytic psychotherapist in the North of England.

Chris Goodey formerly taught at Ruskin College, Oxford, and at the Open University; he is currently on the staff of the Social Science Research Unit at the University of London.

Josephine Green is a research psychologist who has carried out substantial research into people's experiences of prenatal testing and of genetic disorders. From 1985 to 1996 she was Senior Research Associate at the Centre for Family Research, University of Cambridge. She is now Senior Lecturer in Midwifery Studies and a member of the Centre for Reproduction, Growth and Development at the University of Leeds.

Trefor Jenkins is Professor of Human Genetics, School of Pathology, South African Institute for Medical Research, University of the Witwatersrand, Johannesburg, Republic of South Africa.

Gulsan Karbani is a genetic counsellor at St James' University Hospital, Leeds. She has a particular interest in counselling issues.

Jennifer Kromberg is Associate Professor in the Department of Human Genetics, School of Pathology, South African Institute for Medical Research, University of the Witwatersrand, Johannesburg, Republic of South Africa.

Helen Macbeth is Principal Lecturer in Biological Anthropology, Oxford Brookes University, Oxford, and Chair of the Biosocial Society.

Bernadette Modell is Professor of Community Genetics in the Department of Primary Care and Population Sciences, University College London Medical School, London.

Robert Mueller is Consultant Clinical Geneticist at the St James' University Hospital in Leeds.

Evelyn Parsons, a medical sociologist, is Senior Lecturer in the Department of Nursing Studies and Research Fellow in Medical Genetics at the University of Wales College of Medicine in Cardiff.

Frances Price is Senior Research Associate at the Centre for Family Research, Social and Political Sciences Faculty, University of Cambridge.

Sue Proctor is Senior Lecturer in Midwifery at St George's Medical School, London. Until recently she was Research Fellow at the Clinical Epidemiology Research Unit, Bradford Health Authority.

Nadeem Qureshi is a general medical practitioner and Clinical Lecturer in General Practice at the University of Nottingham.

Martin Richards is Professor of Family Research at the Centre for Family Research, Social and Political Sciences Faculty, University of Cambridge. The Centre has a programme of research on the psychosocial consequences of the new genetics.

Ursula Sharma is Professor in Sociology at the School of Education and Social Science, University of Derby. She has a training in anthropology and sociology.

Iain Smith is Senior Lecturer in Health Services Research and Honorary Consultant in Child Health at the Nuffield Institute for Health, University of Leeds.

Pat Spallone is Lecturer at the Centre for Women's Studies, University of York. Her interests include the new reproductive technologies and genetic engineering.

Meg Stacey is Emeritus Professor of Sociology at the University of Warwick.

Marilyn Strathern is Professor of Social Anthropology at the University of Cambridge.

Introduction
Angus Clarke

I

CULTURAL BARRIERS

The existence of different cultures implies that they are distinct and separable, and hence implies the existence of barriers between them. While cultural differences are being eroded by travel, commercial factors and the media, real differences do exist between the lives and experiences of people living in different communities. If we define cultural barriers as those obstacles to mutual understanding that arise from underlying differences in how the world is seen and experienced by people living in different communities, then it is clear that the barriers are not monolithic entities determined by the setting of one's birth but may be fluid, shifting with the changes in an individual's biography.

It is apparent that cultural differences between social groups can develop for many reasons, and are not associated solely with 'race' or ethnicity but with a wide range of formative experiences. Such differences reflect differences in perspective – different ways of understanding the world that have arisen as the result of different experiences. To categorise individuals simply on the basis of their 'race' is to employ stereotypes that restrict and impoverish one's understanding of other people. Indeed, as will be argued at greater length in Chapters 3 and 4, the notions of 'race' or ethnicity are not fixed, objective, biological *facts* about an individual or group; they are shifting assertions that depend upon the context for their validity (Senior & Bhopal 1994, Bradby 1995). In a book such as this – devoted to cultural differences in the context of genetics – it would be all too easy to slip into stereotypes. We hope this tendency has been largely avoided. For a wider discussion of the importance of 'race' in health care in Britain see Ahmad (1993).

In the course of the meeting from which this book has developed, it became clear that there were as many cultural barriers between the different professional groups present at the meeting as there are between the different ethnic groups in British society. Thus, health

1

professionals can feel unfairly criticised and threatened by social scientists who are interested in examining the moral and political assumptions implicit in their daily work. Social scientists, in turn, can feel ignored and undervalued by the medical profession, and denied access to proper research funds and facilities unless they are willing to compromise their independence and to grant control over their research to the clinicians whose practice they should be observing and interpreting. Psychologists occupy a privileged middle ground because they can function in a clinical role in their own right, and they have a battery of clinical investigations that are also research tools. Perhaps, too, they indulge in less interpretation of their subjects' behaviours than do sociologists. Clinicians have therefore often felt more comfortable (less threatened) in research collaborations with psychologists than with sociologists.

In addition to differences in 'race' and professional background, there are many other cultural differences that may be apparent in the genetic counselling clinic. These differences will include the distinction between lay and professional, and will often include distinctions in social class and gender. Differences in linguistic competence may constitute a very important cultural barrier for those clients who are not offered genetic counselling in their mother tongue. These other cultural differences must not be forgotten. In considering the outcomes of pregnancies in women from different ethnic groups, for example, it is most important to allow for the confounding effects of social factors such as income, education and class (Bagley 1995, and see Chapters 6 and 17).

THE USES OF 'CULTURAL DIFFERENCE' WITHIN HEALTH CARE

Before moving on to the next section of this introduction, it may be helpful to consider the ways in which cultural stereotypes are actually employed in the health care system. This is written from my own perspective as an insider studying and working in the National Health Service in Britain for 20 years, and is therefore highly personal. It may be regarded as little more than a few anecdotes strung together.

Health professionals seize upon 'cultural difference' to explain those behaviours of their patients that they otherwise find troubling. It may indeed be troubling for doctors to have their advice or instructions ignored or rejected by clients. In this context, stereotyping may serve a useful function, although at a price. If a doctor or nurse is puzzled, irritated or angered by the incomprehensible action of their patient, then to think 'Oh, s/he is a —— [member of some different social group], so what do you expect?' may soothe the irritation by providing a partial explanation for the patient's behaviour. The 'price' of this professional catharsis is twofold: the professional never learns to understand why the patient has behaved as s/he did, and the patient may miss out on a treatment that could be helpful if it was modified to suit their circumstances more closely.

In the context of genetic counselling, there may be many reasons why a family makes reproductive decisions that differ from those expected by the counsellor or other health professionals. In some communities, there may be more confidence in the practical and emotional support provided by family and neighbours for the parents of a handicapped child than is usual in mainstream British society; there may be less fear that family and neighbours will blame the parents for having had such a child. Also, there may be less expectation of 'every child a perfect child' in communities where many parents can remember the high infant mortality and morbidity of South Asia or Africa. For a discussion of the social factors relating to the birth of a child with Down's syndrome in Britain, see Boston (1994); for a discussion of the experience of being deaf and how this can be accepted very positively, see Sacks (1991).

The midwife or obstetrician who notes that Mrs X has *refused* antenatal screening, or has *refused* a termination of pregnancy for foetal abnormality, may regard the decision as regrettable but explicable because 'after all, she is a ——'. In fact, the reasons why Mrs X has *declined* the termination may be much broader than her theology. Religion is not a neat compartment insulated from the rest of life. If she has the self-confidence to do so (Bowes & Domokos 1996), Mrs X may express the reason for her decision – she may justify herself – in terms of her religion. The full explanation for the difference in values between the Western professionals and herself, however, may be much more complex than this; it may encompass a whole array of social attitudes and expectations as well as theological beliefs.

One recent medical paper from North America, entitled 'Cultural and psychosocial considerations in screening for thalassemia in the Southeast Asian refugee population' (Nidorf & Ngo 1993), identifies beliefs that may function as cultural barriers to prenatal diagnosis in that population; the first two are: '1) the religious conviction that abortion is sinful . . ., and 2) prenatal testing . . . will disturb the natural harmony between man and nature'. It is assumed in the paper that these arguments have no validity, and that attempts should be made gradually to break down the resistance to genetic testing for these conditions. The reasons given for promoting antenatal diagnosis are that it would help to reduce the incidence of thalassaemia and the cost of medical care for the community of refugees. Although the paper contains a sensitive account of many problems experienced by the refugees, its main thrust is instrumental – how to spread the 'benefits' of Western genetics to this community. Such a paper is, therefore, operating in the same judgemental fashion as the midwife or obstetrician in the preceding paragraph.

White women or couples who *decline/refuse* antenatal screening or terminations of pregnancy for foetal abnormality can also feel the disapproval of health service personnel. If an educated, middle-class couple decline testing, it may be made clear to them – perhaps subtly, perhaps not – that they are behaving irresponsibly. The powerful, but crucially ambiguous question, 'Don't you want to make sure your baby is alright?' may be put to them. In such a circumstance, the woman/couple may be exempted from disapproval if they make an appeal to their religion. This appeal to religion is, in effect, a disarming gesture that is intended to placate the professional. I am uncertain about the theological logic – whether adherence to an abstract theological assertion such as belief in God precedes the reluctance to have an abortion, or vice versa – but that nicety can be ignored in this context. The social function of claiming a religious conviction is clear; it is a way of stating, 'My attitude is fixed and will not be shifted by argument, so please stop harassing me.'

Both professionals and patients, then, can resort to the use of stereotyping cultural labels in their interactions. The professionals may be seeking to explain their patient's peculiarities or to soothe their own wounded dignity, while the patient may be seeking to minimise disapproval or pressure from the professionals.

II

GENETIC DISEASE AND INHERITANCE

Our starting point must be genetic disease. Many diseases have been observed to cluster within families, and such observations can be explained by genetic or environmental factors or by a combination of both. For example, several people in a household may acquire the same intestinal infection if their water source is contaminated, or they may develop similar cardiac problems if their diet is too rich and their lives are too sedentary; if they share a genetic susceptibility to such problems then the familial clustering of the disease will be even more marked.

Only a modest proportion of the overall burden of disease is caused by simple genetic factors. While there is a large number of diseases that are determined by mutations in single genes, these are individually rare. Other disorders have predominantly environmental causes, such as trauma, although the 'environment' in this context must be taken to include social and political structures that influence, for example, government transport policy and hence the frequency of road traffic accidents. Most common disease, however, arises from an interaction between genetic susceptibility and environmental variables; these diseases are multifactorial in origin.

The contribution of genetic factors to the entire range of diseases is increasingly being recognised. This is partly because an increasing proportion of health problems have a clearly genetic basis; the previously high mortality from infectious diseases in Britain, especially in childhood, has fallen steadily over the past century, so that a much greater proportion of disease in childhood is now genetic in origin. It is also because our understanding of the genetic contribution to disease has grown enormously in the past 10–20 years.

There have been enormous advances in our understanding of the single-gene (Mendelian) disorders over the past 15–20 years. In contrast, the genetic dissection of the common multifactorial diseases is in its early stages; it is still not sufficiently advanced to have had much impact upon health care or genetic counselling. This situation is certain to change, but when we discuss genetic disease in this volume we are referring to those diseases caused predominantly by single genetic factors.

WHAT ARE GENES?

Genes are coded messages that control the growth, development and function of the body. A person generally has two copies of each gene, receiving one copy of each in the sperm from their father and the other copy in the egg cell from their mother. In relation to disease, genes fall into two broad categories: those for which both copies must function normally for the individual to avoid having a disease, and those for which only a single intact copy is required.

If a person has one faulty (mutated, altered, defective) copy of a gene for which two intact copies are required, that individual will be likely to have or to develop a disease. The faulty copy of the gene is likely to manifest, and the mutated gene is described as being 'dominant' over the normal copy. An individual with such a dominant genetic disorder may transmit the same condition to any child, the chance of the child also carrying the same faulty copy of the gene being 50 per cent (or 1 in 2). The parent will pass on, at random, either the intact copy of the gene or the faulty copy.

If someone has one faulty copy of a gene for which only one functioning copy is required, they will have no problem; such a person is a healthy carrier of a mutated gene. The average person is thought to carry a single mutated copy of several such 'recessive' (hidden) faulty genes. The chance that a carrier of a recessive gene will transmit their faulty copy of the gene to a child is again 50 per cent, and such a child will also be perfectly healthy unless they receive another faulty copy of the same gene from their other parent. If both parents are carriers for the same recessive disorder, the chance of a child having a double dose of the faulty gene, and therefore being affected, is 25 per cent (or 1 in 4).

For recessive disorders, the chance of a child being affected depends crucially upon the frequency of the faulty copy (the mutant allele) of the relevant gene in the population. While having two faulty copies (alleles) of such a gene is clearly disadvantageous, the healthy carriers with a single faulty copy of the gene may have a considerable biological advantage over fully 'normal' individuals; that is why these recessive gene problems have become so common in certain populations. For example, many of the recessive haemoglobin disorders, such as sickle cell anaemia and thalassaemia, are most common in populations which have been exposed over many generations to malaria. Despite the morbidity and mortality suffered by those with the disease (with two faulty copies of the relevant haemoglobin gene), the frequency of such faulty genes has remained high because healthy carriers with a single altered copy of the gene are protected against the worst effects of certain malarial infections.

In any population, there are many more healthy carriers of such a recessive gene than there are affected individuals with a double dose. For cystic fibrosis, for example, there is approximately one affected child for every 2,500 infants born in Britain, but about one person in 25 is a healthy carrier.

FACTORS INFLUENCING THE FREQUENCY OF GENETIC DISEASE

At this point, I must emphasise two important issues that will reappear frequently in this volume, and which anchor the social context of genetic disease within a biological frame.

First, the number of individuals in a population affected by a recessive disorder will depend upon the frequency of the disease allele (the 'faulty' copy of the gene) in the population. There are very substantial differences in the frequency of such genes between different ethnic groups and geographical populations; accordingly, recessive disorders vary greatly in frequency between populations. This is largely a result of the biological history of each population, and reflects the processes of natural selection that have been important until at least the recent past.

Second, the number of individuals affected by recessive disorders in a population will also depend upon the pattern of marriage practised within the community. For any given frequency of the faulty gene, the number of affected children born will be greater if marriage partners are likely to be related to each other 'by blood' (to use a British metaphor): i.e. if they are likely to share some genes inherited from a recent common ancestor or ancestors. This is termed consanguinity ('the coming together of blood').

'FAULTY GENES'

I have been using the words 'normal' and 'faulty', but it is clear that these words can be regarded as value-laden and prejudiced, even discriminatory. It can always be objected that 'normal' is being used in a neutral, statistical sense, but the potential for misunderstanding is clear. In using the word 'faulty' in this introduction, I am not intending to discriminate against people who may carry such genes, but I am indicating, in an impersonal, biological sense, that a particular copy of a gene does not function adequately. In discussing a genetic disease with a family, I would sometimes use the word 'faulty' if I had first established that the family did regard the disease in question as a serious problem. The families of children affected by a serious disorder – progressive, disabling and perhaps fatal – will very often be the first to agree with this evaluation of the disease gene. Euphemisms such as 'altered' or 'unusual' may be appropriate in discussions with some families, but the word 'faulty' will often seem more natural when the family is familiar with the full impact of a serious disease.

This discussion around the use of the term 'faulty' in relation to genes is not intended to suggest that one can always clearly categorise genes as 'good' or 'bad'; individuals may differ in the judgements they make about the same gene. There is no formula that can ensure that words are always used appropriately, and in genetic counselling it is necessary to develop tact, sensitivity and a willingness to find out what a family thinks and feels about the disease under consideration.

GENETIC COUNSELLING

What is genetic counselling? What does it achieve? These questions are difficult to answer in just a few words. One starting point is to consider the types of questions that may be brought to the genetic counsellor by clients. Typical questions include: What is the condition that affects [name of family member]? Why (how) did this condition come to happen in my family/child? How is it caused? How will it progress? Is there any effective treatment? Might anyone else in the family be affected in the future? If so, will they be affected in the same way or might it affect them differently? Is there any way of preventing the condition from developing? If it leads on to complications, can any of these be prevented or treated? If the condition cannot be 'put right', then how can it best be coped with? Is it possible to test to see who may be at risk of it, or at risk of having affected children? Can pregnancies be tested to identify those who would be affected?

A concise and very useful definition of genetic counselling has been provided by Harper (1988): 'Genetic counselling is the process by which patients or relatives at risk of a disorder that may be hereditary are advised of the consequences of the disorder, the probability of developing and transmitting it and of the ways in which this may be prevented or ameliorated.'

Genetic counselling may also be defined as what happens when an individual, a couple or a family asks questions of a genetic counsellor about a medical condition or disease that is, or may be, genetic in origin (Clarke 1994). This definition focuses on the process of genetic counselling, and leads on to a description of what is involved in the process: listening to the client's anxieties and questions, establishing or confirming the diagnosis of the relevant condition in the family, and then answering the client's questions as far as possible by communicating the relevant information to them. This process extends to helping clients to think through the implications of the information, and helping them to consider in advance the consequences of any decisions they make about reproduction, lifestyle or genetic testing; this may be termed scenario decision counselling. Finally, the process includes the provision of support for those affected by a genetic disorder or at risk of it, or with an affected child. This support may be essentially moral, or it may encompass medical surveillance for complications of the disorder.

From these definitions, it can be seen that genetic counselling consists largely of listening and talking. It is initiated by the client, who seeks information or explanations from the counsellor, and is a very individualised process. The ethos of genetic counselling is for the professional to respond to the client's concerns and not to impose his or her own understanding or solutions to the problem. The ways in which clients understand the information provided by the geneticist and the ways in which they may utilise it in arriving at decisions depend upon many factors – including their previous experiences, their prior level of knowledge and understanding, their attitudes, beliefs, values and social obligations, and their sense of personal identity.

In relation to decisions with which the client may be faced, the professional may help them to think through the consequences of the possible options open to them, but he or she will not attempt to lead the client to make one decision rather than another; it is, after all, the clients who will have to live with the consequences of their decisions. The counsellor will therefore attempt to be completely non-directive, even if this goal is unattainable (Clarke 1991).

The definition of the goals of genetic counselling is important, and may be reflected in the ethos of a clinical genetics service as it functions in daily practice. It is, however, difficult and even contentious, and can be approached very differently by those from different disciplines. A public health physician or health economist may have very different views on the important outcomes of genetic counselling from those held by genetic counsellors, and these differences in perspective may lead to important differences in how a genetics service operates, and therefore in how it is experienced by clients from any culture or community (Clarke et al. 1996).

GENETIC SCREENING

Genetic screening is very different from genetic counselling. Genetic screening applies to a whole population or to a distinct subgroup within a population, such as newborn infants or pregnant women; in contrast, genetic counselling applies to individuals or, at most, to small family units. Genetic screening entails the application of a test to a population – the screening is inseparable from the testing; genetic counselling may or may not lead on to genetic testing. In essence, genetic counselling is a small-scale affair in which individuals or families come to professionals seeking information and the professionals respond as best they can. Genetic screening deals with large groups rather than individuals, and it is the professionals who advertise the service and even approach the potential clients to offer them the test. The professionals are being proactive rather than reactive.

The pattern of health service provision of genetic counselling services, and of prenatal screening services for genetic conditions, has evolved in a fairly haphazard fashion; new technologies and new tests have often been introduced as they have become available because of pressure from service providers. Only subsequently has 'public demand' for the services become a significant factor in ensuring their widespread provision, so that even this 'public demand' can be seen as profession-led. Because these developments have taken place in the context of the dominant, orthodox medical–scientific culture, relatively little attention has been paid to the impact of these developments upon the service users. This is now being rectified, but the psychosocial studies of genetics services in Britain that have appeared so far have been largely blind to the ethnic and cultural diversity of the service users.

There are four contexts in which genetic screening can be made available – newborn screening, antenatal screening, carrier screening and susceptibility screening. Screening for complications of genetic disease, such as screening for tumours in those known to be at high risk of malignancies because of a family cancer condition, may also be regarded as genetic screening but is a very different activity from genetic screening as defined here.

Newborn Screening

Newborn screening is carried out in Britain on virtually all newborn infants, and the conditions being screened for are phenylketonuria (PKU) and congenital hypothyroidism; these are preventable causes of severe intellectual disability. If they are diagnosed early and if treatment is started promptly, then the outlook for the affected children is greatly improved – indeed, essentially normal. As such, newborn screening is generally non-controversial; it is the 'acceptable face' of genetic screening. Despite this, there are some issues that do need to be examined in relation to newborn screening.

First, screening for serious but currently untreatable disorders, such as Duchenne muscular dystrophy (DMD), would not fulfil the standard criteria of a worthwhile screening programme because there is no effective treatment for affected but presymptomatic infants. Screening for such conditions, however, may be justifiable if families find the early diagnosis to be helpful in other ways. Possible advantages include the offer of genetic counselling before any further pregnancies, the avoidance of distressing diagnostic delays and the ability to plan realistically for the future ('forewarned is forearmed'). These potential advantages may be offset – in at least some families – by the distress of an early diagnosis spoiling the first few years of the child's life ('ignorance is bliss'). A social evaluation of newborn screening for DMD is in progress in Wales (Bradley et al. 1993, Parsons and Bradley 1994, Parsons et al. 1994, Fenton-May et al. 1994, Parsons et al. 1996)).

Second, there is the question of informed consent for newborn screening (Faden et al. 1982). Should this be required for such programmes? For the essentially untreatable diseases like DMD, parental consent is clearly required before tests are carried out. But for diseases like PKU, where the affected child benefits enormously from early diagnosis, should informed parental consent be required before testing? Or should the tests be mandatory, as in some parts of the USA? Practice varies widely and there is no general consensus, although most observers would agree that the current general ignorance among parents about even the standard newborn screening tests is disappointing (Statham et al. 1993), and the information made available to parents about these tests should certainly be improved.

Antenatal Screening

Antenatal screening programmes are intended to identify pregnancies in which the foetus has a serious malformation such as spina bifida, or a genetic condition such as Down's syndrome (trisomy-21), so that the pregnancy can be terminated if the pregnant woman so decides. Antenatal screening is offered routinely to most pregnant women in Britain, many of whom are 'led' to have such tests without being encouraged to give thought in advance to the underlying issues (e.g. 'Would I want to terminate this pregnancy if the foetus was found to have Down's syndrome or spina bifida?').

The tests most widely available are maternal serum screening (not itself a diagnostic test, but modifying the chance of the foetus having Down's syndrome when interpreted in conjunction with other information, such as the woman's age) and ultrasound scanning to detect malformations such as spina bifida. If ultrasound scans indicate that a chromosomal anomaly may be present, or if maternal serum screening indicates a probability of Down's syndrome occurring in that pregnancy of at least 1 in 200–250, then the obstetrician is likely to suggest an amniocentesis. Chromosome analysis will then be carried out on cells from the fluid surrounding the foetus in the womb. This procedure has a risk of causing miscarriage frequently quoted as 1 in 150 although the figure actually varies widely between units. If the chromosome test identifies an unusual set of chromosomes, the pregnancy is likely to be terminated.

Obstetric and midwifery staff frequently offer the screening tests in such a way that it is difficult for pregnant women to refuse (Sjogren & Uddenberg 1988, Clarke 1991), and in a manner that makes it clear that the woman is expected to comply with the offer (Marteau, Slack et al. 1992, Marteau et al., 1993). Antenatal screening has become routinised at least partly because the possible consequences of compliance or participation – being confronted with a decision about terminating a wanted pregnancy – is glossed over (Press & Browner 1994, Browner & Press 1995). This can be seen as a strategy on the part of professionals, and perhaps pregnant women, to minimise their sense of awkwardness and anxiety at discussing such difficult topics. At the same time, the offer of antenatal screening tests to so many women may have fundamentally altered the experience of being pregnant in Western society; women have come to regard pregnancy as a provisional – 'tentative' – state until the foetus has passed the various screening checks (Rothman 1988).

Advocates of antenatal screening sometimes justify these programmes on the grounds that they will prevent the suffering of the
children who would have been born, or because they place a high
value on the informed decision-making of pregnant women. The most
influential type of justification, however, and the argument that has
been most widely employed in medical journals, has probably been the
cost-benefit argument that it is cheaper to screen pregnancies for
foetuses with certain genetic conditions and to terminate the affected
pregnancies than it is to provide a reasonable level of care for affected
individuals. Variations of this argument have been circulating in Western society for at least a century and in respectable medical journals in
Britain ever since the legalisation of abortion (Hagard & Carter 1976,
Wald et al. 1992). There are good grounds for contesting these arguments, including the concern that the cost-benefit arguments could
drive clinical practice inappropriately towards screening programmes that are still more directive and coercive than at present (Clarke
1990).

Carrier Screening

Carrier screening sets out to identify the unaffected, healthy carriers of
certain recessive genetic disorders. As explained above, these conditions differ in frequency between different ethnic groups, so that the
conditions for which carrier screening may be most appropriate will
differ similarly (Bradby 1996). There are clearly opportunities for
racial discrimination to exploit differences in the frequency of genetic
disease between racial groups, as happened with sickle cell screening in
the USA during the 1970s (Hampton et al. 1974). Even within communities, it is possible for healthy carriers to suffer stigmatisation and
discrimination (Stamatoyannopoulos 1974; Evers-Kiebooms et al.
1994). Furthermore, those identified as carriers can harbour continuing regrets or worries (Zeesman et al. 1984), may feel less optimistic
about their own future health than non-carriers (Marteau, van Duijn
et al. 1992), and may in any case fail to recall the significance of their
test result over time (Loader et al. 1991, Axworthy et al. 1996).

Individuals who are identified as carriers of serious recessive disorders may find this identification useful in making their reproductive
plans – if they are sufficiently 'objective' about such personal matters.
There is also a potential cost to such knowledge, however, in terms of
distress and confusion – as outlined above. Many more individuals
will be identified as carriers in a screening programme than will be at

risk of having a child affected by the disease in question, so these effects are important. Furthermore, experience suggests that many individuals will comply with the suggestion that they have screening for such conditions even though they have little motivation or interest in being screened – the rate of uptake of screening tests for cystic fibrosis varies enormously with the mode of invitation (Bekker et al. 1993, Tambor et al. 1994). It would probably be simple for a commercial offer of carrier screening to be promoted widely and effectively, with many carriers being identified who had not seriously considered the potential disadvantages of testing in advance and who then might regret having had the test. Even in a state health sector, it may be important to examine the interests of those professionals who seek to establish carrier screening: in whose interests would such a programme operate (Koch & Stemerding 1994)?

The identification of carriers of recessive diseases raises social as well as personal issues. How will a community respond to the identification of carriers for such diseases? Will the carriers be stigmatised and regarded by non-carriers as unhealthy or tainted, and therefore as unsuitable partners in relationships? Will racial discrimination be exacerbated by discrimination against the carriers of a recessive disease which is found predominantly in one ethnic group? If a carrier has an affected child, will they be held accountable? (Duster 1990)

Susceptibility Screening

Traditional genetic studies of human diseases have concentrated upon those unusual families in which a single gene is clearly responsible for a particular disease. Much has been learned in this way, but these studies have had little impact upon the understanding or management of the more common, everyday diseases such as bowel cancer, breast cancer, heart disease, diabetes, hypertension and dementia. In the last few years, the molecular genetic study of these multifactorial diseases – in which environmental factors, chance and genetic factors interact to cause ill health – has begun to dissect out the contributing genetic factors.

These developments may well be helpful in the long term if they lead to improved treatments, but in the short term this is unlikely. Technology will gain the power to identify those at risk of disease without preventing it or improving their treatment. The current screening test that is closest to genetic susceptibility screening is probably cholesterol screening, which is recognised as causing problems as well as leading to benefits for a few. Newborn screening for alpha$_1$-antitrypsin deficiency – a condition that can lead to lung disease in those exposed to smoke or dust – paradoxically resulted in potentially harmful effects in the families of the identified infants; in addition to great anxiety, the fathers of affected children were found to smoke more than in control families (McNeil et al. 1988). Benefits from lifestyle changes are in any case likely to be modest and could probably be gained without the expense of genetic testing (Clarke 1995). The identification of those at increased risk for common diseases has the potential to cause distress, denial and inappropriate feelings of fatalism, and opens up the possibility of discrimination against those at risk in relation to life and health insurance, employment and other spheres of life (Davison et al. 1994; Clarke 1995).

How will society use the new capacities to identify those at increased risk of disease? If a genetic basis for an increased risk of heart disease or diabetes, for example, is found in one ethnic minority group within Britain, will this be exploited by racists to justify institutionalised discrimination against that ethnic group in relation to immigration, employment or insurance? Will members of the ethnic minority group take a positive view of their 'risky' genes by adopting and promoting lifestyle changes that could counteract the risk factors?

The possible implications of the genetic dissection of other multifactorial traits not necessarily related to a disease, such as intelligence and personality, are discussed elsewhere, and may be of interest to readers of this volume because of the potential misuse of such information to promote racist beliefs (Clarke 1995).

III

THE SOCIAL SCIENCES AND 'CROSS-CULTURAL GENETICS'

Why are social studies of genetic counselling and screening important? Why should they pay attention to ethnic and cultural diversity?

Genetic counselling consists largely of talking; genetic counsellors deal in information and explanation while other health professionals prescribe pills or wield the knife. Once the relevant medical diagnosis has been established as far as is possible, the predominant activity of the genetic counsellor is talk. The client will then respond to the information provided, and will respond in a broader sense to the process of genetic counselling and not just to the information content. Many of the factors likely to influence the client's response to genetic counselling will vary enormously from one person or family to another. Individuals and families from the same ethnic, cultural or religious group may share some of these factors and this may lead to common patterns of response. This assumption, however, is problematic in that it carries implicit notions of cultural, ethnic and religious groups and of their influence on behaviour. There is a need to steer between the Scylla of stereotypes and the Charybdis of radical, 'insular' individualism. Such terms as 'culture' and 'ethnicity' must not be reified, treated as if they describe concrete entities, when they are in fact only loosely shared mental categories, but it is also important not to err in the other direction by denying the reality of social factors and thereby concentrating exclusively upon individuals while ignoring the social context in which the individuals lead their lives. It would be easy for the individualist perspective to be adopted by default in contemporary Western society, but this would justifiably be viewed as a form of ethnocentrism from the perspective of other 'cultures'.

Psychosocial studies of genetic counselling have a short history, and have so far generally avoided the issues of race, religion and other difficult areas of 'culture', although there have been some exceptions (e.g. Modell 1991). From the perspective of clinical genetics in Britain, however, it is important that these areas are examined. As offers of genetic counselling and testing become more frequent, more clients from ethnic minority groups in Britain are interacting with clinical genetics services. In particular, clients/patients from specific ethnic groups may be offered genetic screening tests for conditions whose prevalence is greater in these groups than in Britain as a whole, so that awareness of biological links between ethnicity and disease is increasing.

In addition to the problems experienced by individuals from any ethnic group when they are in contact with genetics services, clients from minority groups may also have linguistic barriers to overcome and there may be other 'cultural differences' that influence their responses to the genetic counselling or screening. Consequently, any psychosocial difficulties arising in the course of genetic screening or counselling are likely to be amplified when ethnic minority groups are involved with these services. It is therefore important that the ethnic and cultural dimension to genetics services is given careful attention.

From the perspective of service providers, this will enable us to improve the service offered to all clients, from whatever background.

But why should social scientists be interested in investigating this area?

First, studies in this area may illuminate a more general set of questions about the relation between belief and action – between systems of belief and patterns of behaviour. In particular, studies of cross-cultural genetic counselling may clarify the relation between concepts of inheritance, disease and morality on the one hand, and aspects of behaviour and decision-making on the other. The relevant decisions may relate to reproductive decisions, the sharing of personal genetic information with other members of the family, the approach to coping with disease and disability, and the means used to obtain personal support at difficult times.

Second, there is an opportunity for social scientists to improve their understanding of popular ideas of inheritance and disease. Not only are these topics brought up for discussion in genetic counselling, but there are also some defined endpoints and outcomes that can be measured objectively. These include the making of reproductive decisions, the diffusion of genetic information within a family and the performance of diagnostic tests. It may therefore be possible to gather statistically useful data in these studies, which could greatly facilitate the testing of explanatory hypotheses. The uptake of genetic counselling, screening and testing, and the uptake of health screening in those thought to be at increased genetic risk of health problems, can be regarded as measures that inform us about the relationship between understanding, beliefs and values on the one hand and real-life decisions and behaviours on the other. A tentative exploration of individuals' beliefs in this area has been reported by Ponder et al. (1996).

Third, there is the idea developed forcefully by social anthropologists that an understanding of a different society illuminates one's own. For example, it is only because other kinship patterns have been studied by social anthropologists that it has been possible for Strathern to study contemporary English kinship from such a detached perspective (Strathern 1992).

Finally, there is another area of interest to social scientists which could be illuminated by cross-cultural studies of genetics services. The coexistence of several different 'cultures', with one conventionally regarded as the 'dominant', 'mainstream' culture, raises questions about communication and convergence between these cultures. Studies

of genetic counselling and screening, and their influence on clients'
ideas of inheritance and genetic disease, may illuminate the more
general pattern of relationships between coexisting cultures, including
the processes of cultural assimilation and cultural change.

IV

OUTLINE OF CONTENTS OF THIS BOOK

Part I consists of four essays that provide the conceptual framework for
the rest of the volume. In the first essay, Bernadette Modell describes
the kinship patterns practised among some of the ethnic minority
groups in Britain today. With her clinical experience of families with
affected children, she discusses the social implications of these different
kinship patterns with particular attention to the consequences for the
status of women. Marilyn Strathern's paper, very much as it was
delivered at Abergavenny, is more abstract. She elucidates the concepts
of 'culture' and 'relatedness', emphasising how we are always more
aware of other cultures than we are of our own and that we can learn
much about the concept of kinship within Western society by attending
to the cultural implications of difficult scenarios in genetic counselling.
Throughout the meeting in Abergavenny, there was a constant return to
the themes elaborated by both Modell and Strathern.

Helen Macbeth and Ursula Sharma then consider the question,
'What is an Ethnic Group?' Macbeth writes from a biological perspective, and explains clearly the biological meaning of this term. This
amounts essentially to explaining what it does *not* mean. She demonstrates that finding biological differences between two groups of individuals (however chosen) does not justify the conclusion that they
represent different 'races' or distinct 'ethnic groups'; the very idea of
such discrete biological groupings is itself flawed. Sharma discusses
some of the different ways in which the terms 'race' and 'ethnic group'
are used. She is eager to expose the over-simplistic use of these terms
in a variety of contexts including contemporary politics. These two
essays complement each other, and clarify the use of these important
terms throughout the rest of the volume.

Part II contains seven essays that reflect upon experiences of genetic
counselling, or describe empirical studies of genetic counselling in

relation to different ethnic groups. The first three essays refer to consanguinity in British Pakistani communities. Darr (Chapter 5) describes the networks of information and support available to British Pakistani families who have a child with thalassaemia. The consanguinity that may be blamed for the disease in fact assists by establishing wide-ranging networks of kinship support that help families bear the burden of the disease. Proctor and Smith (Chapter 6) examine the claim that consanguinity is responsible for the high rate of adverse birth outcomes observed in British Pakistanis. They demonstrate that there are a number of other factors that contribute to the poor birth outcomes in this group, especially social class and material deprivation. The common medical focus on consanguinity as the dominant explanation for the poor outcomes is misguided, and may reflect and reinforce a subtle form of racism. Qureshi (Chapter 7) reflects upon his experiences as a GP in an area with a number of Muslim families. He examines the ways in which information about inherited diseases can most usefully be disseminated among such families. He castigates those health professionals who blame families for having children affected by genetic disorders if the parents are consanguineous, and (like Darr) shows how the social network that comes from consanguinity helps to support these families.

The next three essays examine issues relating to families in Africa or of African descent. Green and France-Dawson (Chapter 8) examine the experiences of antenatal screening of women from different ethnic groups in the British West Midlands and demonstrate the complex relationship between different ethnic minority groups in Britain. Akinyanju then provides a fascinating account of the evolution of a genetic counselling service for sickle cell disease in Nigeria. This has evolved in response to the needs of families attending clinic for the treatment of their affected children. The involvement of families in the development of the service has been crucial, and the avoidance of official pressure on those at risk of having affected children has also prevented the project from being discredited. In Chapter 10, Kromberg and Jenkins describe some of the cross-cultural issues – the conflicts between systems of belief – that they have witnessed over many years of work in genetic counselling in South Africa. As they state, their paper does not so much provide answers as indicate the need for further study of the different systems of belief in that great and rapidly changing country.

The final paper in Part II, by Karbani, Godsil and Mueller, is rather different from any of the others. It is an interpretation of the imagery used by clients and counsellors in genetic counselling in Leeds and

Bradford. It focuses on the fantasies that clients and professionals have about people of different 'races' – the images that Westerners may have of 'primitive' black or Asian people. These images may be shared by black and Asian people themselves, and may obstruct the therapeutic effectiveness of counselling in support of people affected (directly or indirectly) by genetic conditions. This chapter is a brave attempt by practitioners to voice what is so often left unspoken. Our mental images of our own and other 'races' may be positive (Torgovnik 1990) or negative (Said 1993), but they need to be confronted if they are preventing the resolution of difficult psychological responses to disease.

Part III consists of two essays on the differences between lay and professional understandings of inheritance in relation to disease. Both Davison and Richards have contributed greatly to the exploration of this area (ethnogenetics?) in British families. The clinical implications of 'false' beliefs about inheritance are beginning to become apparent, and are particularly clear in relation to ideas about the inheritance of breast cancer. For example, false ideas in this area could lead some women at risk of familial breast cancer to regard themselves as not at risk if they inherit their risk from affected relatives through their father.

Part IV, the final section, contains five essays on the social and political context of genetics. Pat Spallone examines, and challenges, the dominant role of genetics in current explanations for phenomena that have social as well as genetic causes. Chris Goodey continues this process, referring in particular to learning difficulties and the way in which society categorises those with such disabilities. Frances Price examines the area of gamete donation and the public debate around this topic. She also examines the issue of race in connection with sperm donation and the attempts to 'match' the ethnicity of social and biological fathers. In the final two chapters, Meg Stacey and Evelyn Parsons examine the wider social issues that have arisen from the impact of genetics on society in the recent past.

It is the contributions in this final part of the volume that are most likely to cause discomfort in medical circles, but it is for that very reason that the essays are so important. There are very real social and political issues to discuss in relation to the development of genetics in health care, and these issues must be confronted and not ignored. Only in that way will society be able to benefit from the applications of genetics without walking into the real problems that lie ahead for the unwary. If such views are ignored or marginalised, then the various genetic dystopias that are imagined and feared are more likely to translate into future reality.

REFERENCES

Ahmad, W.I.U. (ed.) 1993 *'Race' and Health in Contemporary Britain*. Open University Press: Buckingham and Philadelphia.
Axworthy, D., Brock, D.J.H., Bobrow, M., & Marteau, T.M. 1996 Psychological impact of population-based carrier testing for cystic fibrosis: 3-year follow-up. *Lancet* 347: 1443–6.
Bagley, C. 1995 A plea for ignoring race and including insured status in American research reports on social status and medicine. *Social Science and Medicine* 40: 1017–19.
Bekker, H., Modell, M., Denniss, G., Silver, A., Mathew, C., Bobrow, M., & Marteau, T. 1993 Uptake of cystic fibrosis testing in primary care: supply push or demand pull. *BMJ* 306: 1584–6.
Boston, S. 1994 *Too Deep for Tears: Eighteen years after the death of Will, my son*. Pandora: London.
Bowes, A.M., & Domokos, T.M. 1996 Pakistani women and maternity care: raising muted voices. *Sociology of Health and Illness* 18: 45–65.
Bradby, H. 1995 Ethnicity: not a black and white issue. A research note. *Sociology of Health and Illness* 17: 405–17.
Bradby, H. 1996 Genetics and racism, in Marteau, T., & Richards, M. (eds) *The Troubled Helix: Social and psychological implications of the new human genetics*. Cambridge University Press: Cambridge, pp. 295–316.
Bradley, D.M., Parsons, E.P., Clarke, A. 1993 Preliminary experience with newborn screening for Duchenne muscular dystrophy in Wales. *BMJ* 306: 357–60.
Browner, C.H., & Press, N.A. 1995 The normalization of prenatal diagnostic screening, in Ginsburg, F.D., & Rapp, R. (eds) *Conceiving the New World Order: The Global Politics of Reproduction*. University of California Press: Berkeley.
Clarke, A. 1990 Genetics, ethics and audit. *Lancet* 335: 1145–7.
Clarke, A. 1991 Is non-directive genetic counselling possible? *Lancet* 338: 998–1001.
Clarke, A. 1994 Introduction, in Clarke, A. (ed.) *Genetic Counselling: Practice and principles*. Routledge: London and New York, pp. 1–28.
Clarke, A. 1995 Population screening for genetic susceptibility to disease. *BMJ* 311: 35–8.
Clarke, A., Parsons, E.P., & Williams, A. 1996 Outcomes and process in genetic counselling. *Clinical Genetics* 50: 462–9.
Davison, C., Macintyre, S., & Davey Smith, G. 1994 The potential social impact of predictive genetic testing for susceptibility to common chronic diseases: a review and research agenda. *Sociology of Health and Illness* 16 (3): 340–71.
Duster, T. 1990 *Backdoor to Eugenics*. Routledge: New York and London.

Evers-Kiebooms, G., Denayer, L., Welkenhuysen, M., Cassiman, J- J., & Van den Berghe, H. 1994 A stigmatizing effect of the carrier status for cystic fibrosis? *Clinical Genetics* 46: 336–43.

Faden, R., Chwalow, A.J., Holtzman, N.A., & Horn, S.D. 1982 A survey to evaluate parental consent as public policy for neonatal screening. *American Journal of Public Health* 72: 1347–52.

Fenton-May, J., Bradley, D.M., Sibert, J.R., Smith, R., Parsons, E.P., Harper, P.S., & Clarke, A. 1994 Screening for Duchenne muscular dystrophy. *Archives of Diseases of Childhood* 70: 551–2.

Hagard, S., & Carter, F.A. 1976 Preventing the birth of infants with Down's syndrome: a cost-benefit analysis. *BMJ* I: 753–6.

Hampton, M.L., Anderson, J., Lavizzo, B.S., & Bergman, A.B. 1974 Sickle cell 'nondisease': a potentially serious public health problem. *American Journal of Diseases of Children* 128: 58–61.

Harper, P.S. 1988 *Practical Genetic Counselling*, 3rd edition. Wright: London.

Koch, L., & Stemerding, D. 1994 The sociology of entrenchment: a cystic fibrosis test for everyone? *Social Science and Medicine* 39: 1211–20.

Loader, S., Sutera, C.J., Segelman, S.G., Kozyra, A., & Rowley, P.T. 1991 Prenatal hemoglobinopathy screening IV: Follow-up of women at risk for a child with a clinically significant hemoglobinopathy. *American Journal of Human Genetics* 49: 1292–9.

Marteau, T., Slack, J., Kidd, J., & Shaw, R. 1992 Presenting a routine screening test in antenatal care: practice observed. *Public Health* 106: 131–41.

Marteau, T.M., van Duijn, M., & Ellis, I. 1992 Effects of genetic screening on perceptions of health: a pilot study. *Journal of Medical Genetics* 29: 24–6.

Marteau, T.M., Plenicar, M., & Kidd, J. 1993 Obstetricians presenting amniocentesis to pregnant women: practice observed. *Journal of Reproductive and Infant Psychology* 11: 3–10.

McNeil, T.F., Sveger, T., & Thelin, T. 1988 Psychosocial effects of screening for somatic risk: the Swedish alpha-1-antitrypsin experience (editorial). *Thorax* 43: 505–7.

Modell, B. 1991 Social and genetic implications of customary consanguineous marriage among British Pakistanis. Report of a meeting held at the CIBA Foundation on 15 January 1991. *Journal of Medical Genetics* 28: 720–3.

Nidorf, J.F., & Ngo, K-Y. 1993 Cultural and psychosocial considerations in screening for thalassemia in the Southeast Asian refugee population. *American Journal of Medical Genetics* 46: 398–402.

Parsons, E.P., & Bradley, D.M. 1994 Ethical issues in newborn screening for Duchenne muscular dystrophy: the question of informed consent, in Clarke, A. (ed.) *Genetic Counselling: Practice and principles*. Routledge: London and New York, pp. 95–112.

Parsons, E.P., Bradley, D.M., & Clarke, A. 1996 Disclosure of Duchenne muscular dystrophy after newborn screening. *Archives of Disease in Childhood* 74: 550–3.

Parsons, E.P., Clarke, A.J., Bradley, D.M. 1994 'They must have got it wrong': The experiences of families identified by a newborn screening programme for Duchenne muscular dystrophy, in Bloor, M., & Taraborrelli, P. (eds) *Qualitative Studies in Health and Medicine*. Avebury: Aldershot, pp. 43–59.

Ponder, M., Lee, J., Green, J., & Richards, M. 1996 Family history and perceived vulnerability to some common diseases: a study of young people and their parents. *Journal of Medical Genetics*, 33: 485–92.

Press, N.A., & Browner, C.H. 1994 Collective silences, collective fictions: how prenatal diagnostic testing became part of routine prenatal care, in Rothenberg, K.H., & Thomson, E.J. (eds) *Women and Prenatal Testing: Facing the challenges of genetic technology*. Ohio State University Press: Columbus.

Rothman, B.K. 1988 (1986) *The Tentative Pregnancy: Prenatal diagnosis and the future of motherhood*. Pandora: London.

Sacks, O. 1991 *Seeing Voices*, revised edition. Picador: London.

Said, E.W. 1993 *Culture and Imperialism*. Vintage: London.

Senior, P.A., & Bhopal, R. 1994 Ethnicity as a variable in epidemiological research. *BMJ* 309: 327–30.

Sjogren, B., & Uddenberg, N. 1988 Decision making during the prenatal diagnostic procedure. A questionnaire and interview study of 211 women participating in prenatal diagnosis. *Prenatal Diagnosis* 8: 263–73.

Stamatoyannopoulos, G. 1974 Problems of screening and counselling in the hemoglobinopathies, in Motulsky, A.G., & Ebling, F.L.B. (eds) *Birth Defects: Proceedings of the Fourth International Conference*. Excerpta Medica: Amsterdam.

Statham, H., Green, J., & Snowdon, C. 1993 Mothers' consent to newborn screening babies for disease. *BMJ* 306: 858–9.

Strathern, M. 1992 *After Nature: English kinship in the late twentieth century*. Cambridge University Press: Cambridge.

Tambor, E.S., Bernhardt, B.A., Chase, G.A. et al. 1994 Offering cystic fibrosis carrier screening to an HMO population: factors associated with utilization. *American Journal of Human Genetics* 55: 626–37.

Torgovnik, M. 1990 *Gone Primitive: Savage intellects, modern lives*. University of Chicago Press: Chicago and London.

Wald, N.J., Kennard, A., Densem, J.W., Cuckle, H.S., Chard, T., & Butler, L. 1992 Antenatal maternal serum screening for Down's syndrome: results of a demonstration project. *BMJ* 305: 391–4.

Zeesman, S., Clow, C.L., Cartier, L., Scriver, C.R. 1984 A private view of heterozygosity: eight-year follow-up study on carriers of the Tay-Sachs gene detected by high school screening in Montreal. *American Journal of Medical Genetics* 18: 769–78.

Part I
Background

1 Kinship and Medical Genetics: A Clinician's Perspective[1]

Bernadette Modell

This article is written from the point of view of a clinician whose main clinical interest for 30 years has been treatment and prevention of the haemoglobin disorders (Modell & Berdoukas 1984). This has involved regular contacts with people from most of the main ethnic minority groups in the UK, and has permitted some understanding of the way that kinship patterns can affect a family's ability to cope with inherited disease and their use of genetic counselling. The concepts outlined here have developed gradually over a number of years, in the light of dicussions with professional colleagues drawn from a wide range of ethnic groups, particularly Dr Aamra Darr (whose Ph.D. work included some new insights into the social role of customary consanguineous marriage (Darr 1991, Modell & Kuliev 1992)), Drs Anver Kuliev, Nadeem Qureshi, Mary Petrou, Michael Angastiniotis, Elizabeth Anionwu and Sister Quai Peng Wong.

Genetic counselling involves providing people with information, so that they can make decisions that accord with their personal perceptions and moral and social values. In the UK it is now national policy to provide population screening for carriers of haemoglobin disorders (the thalassaemias and sickle cell disorders) (Report 1994), usually during pregnancy. Eleven per cent of births in the UK are to mothers in ethnic groups at risk for haemoglobin disorders, and on average 8 per cent of members of these groups are carriers (Report 1994). Carrier frequency ranges from a high of over 20 per cent among Africans and 17 per cent among Cypriots to about 3 per cent among Indian Punjabis. Carriers are healthy themselves, and because their red blood cells differ from the normal in minor ways, they are protected against death from malignant (falciparum) malaria. This confers a selective advantage in the tropics and sub-tropics, and explains the high frequency of these genes in populations originating from these areas. However, carrier status also confers an important genetic risk. When both of a couple are carriers, they may[2] have a one in four risk of having a child with a major haemoglobin disorder (such as sickle cell anaemia or thalassaemia major) with each pregnancy.

Midwives routinely[3] offer carrier testing to pregnant women in these groups, and when a pregnant carrier is found, her partner is invited for testing. When a couple of carriers is detected, they require expert genetic counselling and the offer of prenatal diagnosis when appropriate, and there is a growing network of trained 'Sickle Cell and Thalassaemia Counsellors' in the UK. Uptake of prenatal diagnosis is very high for thalassaemia, a predictably severe and intractable anaemia that involves very burdensome treatment (Modell et al. 1980). Sickle cell disorders are less predictable, genetic counselling is more complex and the uptake rate of prenatal diagnosis is affected by several other factors, of which the most important is stage of gestation at counselling (Petrou et al. 1992). By the end of 1994, over 800 at risk couples living in the UK had been counselled at the prenatal diagnosis unit at University College Hospital (Modell et al., in press).

The information given to couples at risk for having children with a haemoglobin disorder is basically rather uniform, but it produces a wide range of different reactions that depend on the parents' personality, experiences and cultural setting. In fact, from a research perspective, giving such relatively standard information may be viewed as a 'social probe' (rather analogous to a DNA probe) that can elicit cultural differences in family dynamics. Thus genetic counselling can incidentally be a valuable sociological research tool, with the important advantage that it is not socially invasive, since we are *obliged* to provide genetic information to couples at risk. The experience suggests that we need a basic understanding of the anthropological concepts of kinship if we are to provide sensitive and appropriate genetic counselling for people from a wide range of cultures.

The most important generalisation about migrant ethnic minority groups is that they are in a state of constant change (Watson 1978). Migration to Western countries and immersion in European society naturally affects migrants' social outlook and disturbs traditional kinship structures. One might therefore expect kinship patterns in England to reflect an intermediate stage between the patterns of 'home' and of the host society: however, perceptions of kinship, being integral to individual and family identity, can be highly resistant to change. For example, the rate of first-cousin marriage among British Pakistanis has risen from about 30 per cent among first generation migrants to 55 per cent in the present generation (Darr & Modell 1988). Ethnic differences in kinship pattern will persist for a long time in the UK, and need to be taken into account in planning appropriate medical services.

A number of typical patterns regularly encountered in the UK are outlined below. Though a short description is necessarily over-simplified and crude, awareness of such differences can help health workers to avoid prejudice and to provide appropriate support.

THE INCEST TABOO

Kinship structures (Fox 1966) are maintained by concepts, conventions and powerful sanctions such as the incest taboo which, in almost all cultures, outlaws sexual relations with first-degree relatives (parent, sibling, child). The range of the incest taboo differs with culture. In some it is restricted to first-degree relatives, while in others it excludes sexual relations between more distant relatives, such as uncle and niece, or first cousins.[4] In a limited number it extends to sexual relations with more distant relatives, such as second cousins. The origins of these different conventions are obscure: they usually represent cultural and religious, rather than civil, sanctions.

Most of us take our own society's kinship pattern for granted. We are essentially unaware of it until we notice that other groups behave differently, and then we are inclined to interpret their different behaviour as deviant or abnormal. In Europe population migrations are mixing cultural groups to a previously unforeseen extent, and in view of the power of the incest taboo, differences in marriage customs can contribute to cultural prejudice. For example, the free choice of partner in European societies is seen as verging on the promiscuous by many Asians, while 'arranged' marriage of any type seems intolerably restrictive to most Europeans. Therefore, in discussing kinship patterns it is necessary to start by considering one's own – in my case, a Northern European pattern.

EUROPEAN KINSHIP PATTERNS

Northern Europe

Figure 1.1 shows the typical Northern European kinship pattern, which has been characteristic of Anglo-Saxon society in particular

Primary
kin of A

Primary
kin of B

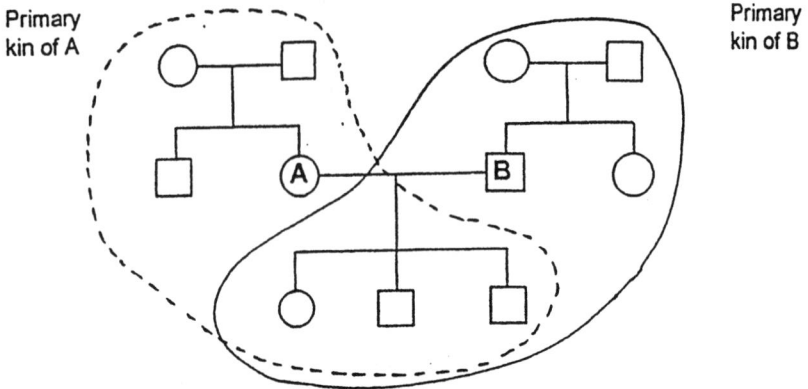

Figure 1.1 Common European ('egocentric') kinship pattern.

for as far back as it can be traced (Fox 1966). Anthropologists call
this an 'egocentric' pattern because all relationships are described
with reference to the person speaking. Marriage to a first cousin
is permitted but discouraged: Catholics require a religious
dispensation, but Protestants do not. Cousin marriage was, and
remains, common in geographically or socially isolated communities
(see many 19th century novels, such as Emily Brontë's *Wuthering
Heights*).

The Northern European pattern corresponds well with genetic reality. Implications are (1) a married couple each have rather different
patterns of kin; (2) kinship pattern is independent of gender; and (3)
relationships established by marriage (other than the relationship of
husband and wife) are tenuous by comparison with blood relationships. Essentially therefore, Northern European society consists of a
large number of small overlapping units (like scales on a butterfly's
wing) that are not very strongly linked to each other in time or
space. Emphasis is on the nuclear family, and outside the main ties
of obligation to first-degree relatives, social relations are with friends
and with society in general. This kinship structure may favour personal development and mobility, social cohesion and individual
enterprise.

The typical Northern European kinship pattern may also have
contributed to the Western origin of the modern commercial society.
Since this in turn strongly promotes the individualism typical of
the 'egocentric' Northern European kinship pattern, changes towards
this kinship pattern now seem inevitable in many other societies.

The Mediterranean

The Greek Cypriot kinship group seems to be an extension of the Northern European one, despite a stronger tradition of protection for women. It has the following characteristics (Constantinides 1978). (1) Marriage is forbidden between relatives as distant as second cousin, and dispensations occur rarely (if at all). (2) At marriage, all kin of both partners view themselves as becoming one family. Consequently most people in this small society are related to each other by marriage or blood, with consequent mutual benefits and responsibilities. (3) Social life revolves around family events such as christenings, marriages and funerals, which typically attract very large numbers of people.

The net result is a society with an unusually democratic atmosphere, and extensive possibilities for family support in case of need. Turkish Cypriot family structures are essentially similar – unsurprisingly, as all available evidence indicates that Greek and Turkish Cypriots are genetically the same people, but with different acquired characteristics. Though first-cousin marriage is permitted among Turkish Cypriots, it is uncommon.

PATRILINEAL KINSHIP PATTERNS

In much of the Indian sub-continent and in the Middle East, North Africa and China, a kindred is defined mainly by the male line of descent from a common ancestor (that is, the kinship structure is patrilineal). Since high value is placed on vertical links, individuals remain associated in large extended families, and people are defined by their place in this family group. A patrilineal kinship pattern is obviously closely related to a tribal concept – members of a tribe consider themselves to be descended from a common male ancestor whose name they bear (Fox 1966) (Figure 1.2).[5] Patrilineal kinship groups also tend to be patrilocal, i.e. males and their descendants stay together, especially when the family owns land. Women leave their family of origin to enter their husband's kindred. The move is often associated with transfer of property, as bride-price or dowry.

A population of this type is made up of many sizeable related groups, which both provide strong support for and exercise considerable control over their members. The numerous family obligations within the

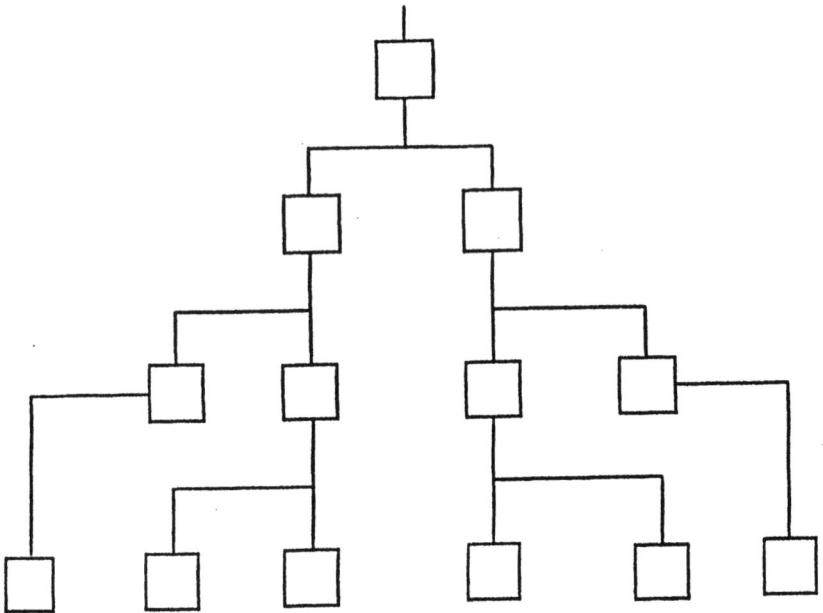

Figure 1.2 Patrilineal kinship structure.

group can limit individual development (for example, in the modern world, by limiting a young person's time and privacy for study). The mutual exclusiveness of such family groupings may be fortified by other social conventions, as in the Indian caste system, and a society composed of such differentiated groups may be difficult to unify.

Patrilineal kinship groups fall into two broad categories depending on where the women come from. These can have very different implications for family relationships, and women's quality of life.

1. *Patrilineal exogamy* (Figure 1.3) is traditionally favoured by, for example, Hindus (higher castes). Women are customarily obtained from unrelated families of the same caste (Moore 1995). At marriage the woman is considered to leave her own family and join that of her husband, a concept that is reinforced by a range of conventions (for example, in some groups a woman cannot offer her parents food in her husband's family home). A dowry is usually paid, the amount depending on the social class relations of the families (Srinivas 1984). A social step up involves a large dowry, a step down little or none at all. Thus, traditionally, girls represent a net burden on their family, a daughter's birth is often greeted with disappointment and dismay, and some parents restrict their emotional investment in a daughter because

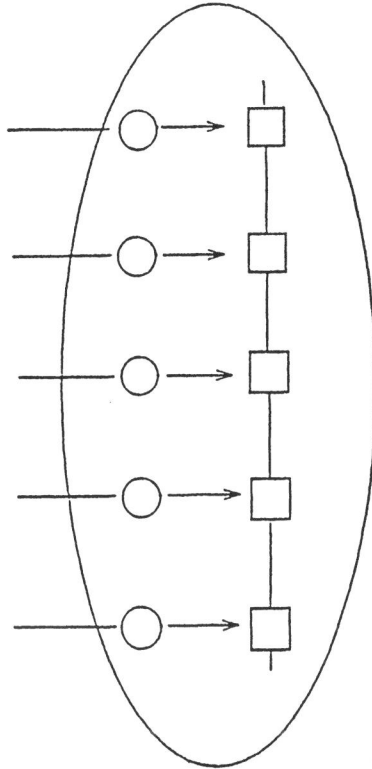

Figure 1.3 Patrilineal exogamic kinship pattern.

they know she is destined to leave them and belong to someone else. Dowry may be perceived as a payment 'to encourage the bridegroom's family to treat the girl nicely'. If the bridegroom's family are dissatisfied with the dowry paid they may treat the woman badly to induce her parents to give more, and can even resort to 'bride-burning' to win a second chance in the dowry lottery (Srinivas 1984, Moore 1995). In India, women's situation is unfortunately getting worse, not better, as commercial development leads to increasing demands for dowry payments (Moore 1995).

Clearly a woman's power position within this family structure is initially very weak. She can gain access to the lines of power and authority only through having a son, and will need to maintain her control over him for as long as possible. When her son marries she will not wish to abdicate in favour of her daughter-in-law. Thus, the forces operating on the women of the family are centrifugal, forcing them into competition, driving them apart and further weakening their power position as a group. I do not mean to say that this is true of every Indian family. Family relations depend primarily on personalities, but when there is competition, the power positions of family members, which are largely determined by kinship pattern, usually determine the outcome.

Reflections of this situation are sometimes seen in the UK, for example in the problems that can arise when a child is found to have a serious inherited disease, such as thalassaemia major. The extent to which the diagnosis affects a family depends primarily on the quality of the relationship between the parents, and most Indian families cope well. However, some extreme adverse reactions can occur among Indian women, some of whom, in my experience, were among the least supported and the most isolated and depressed of all mothers of thalassaemic children. Some felt blamed for bringing sickness into their husband's family, and feared that they would also be held responsible for devaluing their family of origin. Some mothers even hid their child's illness from their own parents or sisters, a reaction I have not observed in any other social group. The only two cases I encountered where parents disagreed about whether an affected child should be treated were both in this group (the mother wanted treatment, the father did not – both (female) children were treated). In both cases it transpired that the father was reflecting pressure from his own mother to abandon the affected child and get a healthy one instead. When I started to work with Aamra Darr (Darr 1991), I was exceedingly surprised to meet a completely different atmosphere among affected British Pakistani families.

2. Pakistanis are among the numerous populations that favour *patrilineal endogamy*, in which women are obtained from within the extended family or tribe (Figure 1.4) – i.e. consanguineous marriage is customary (Bittles 1990). Among Pakistanis (and Indian Muslims) the patrilineal family is surrounded by a larger group of more distant and adoptive relations, the *biradheri*. Most wives are found within this circle, and marriage to a first cousin is preferred. All four possible types of cousin marriage are acceptable. In these circumstances, (1) a daughter is not perceived as a burden because a balanced sex ratio is desired within the extended family, and (2) the commercial aspects of marriage are minimised. As the custom also reduces transfer of members between families, it reinforces the identity of the kinship group · and its distinctness from society in general.

Within such a social structure, genetic disease cannot be viewed as

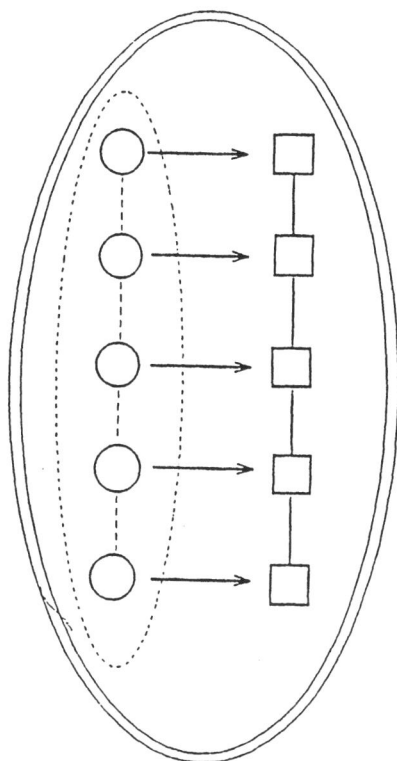

Figure 1.4 Patrilineal endogamic kinship pattern.

stigmatising for the woman, since most family members are blood relatives – the condition is 'in the family'. The Pakistani families with thalassaemic children conveyed no suggestion of blame for the woman and little, if any, feeling of shame about the affected children. There was open discussion of inheritance, prenatal diagnosis etc., and equality between the parents in decision-making on matters such as prenatal diagnosis. It seemed intuitively obvious that there was a relationship between the fundamentally balanced position of the women in these families and the consanguineous kinship pattern.

Aamra Darr pointed out that consanguineous marriage can strengthen the power position of women because when it is common, patrilineal families are also to some extent matrilineal. A woman who marries her cousin has blood ties in her own right with her mother-in-law or father-in-law or both (for example, as her aunt and/or uncle) and is likely to have known them, and her husband, since childhood. Because she has married a relative she often stays in the same village and is not separated from her parents. She can discuss problems with both her in-laws and her parents (at an Arab wedding a father was overheard saying jokingly to the bride, 'If he doesn't treat you right, just you come to me and I'll have a word with his mother'). Thus, the marriage pattern can increase the stability of the entire family and is integral to the support systems of the societies in which it is practised.[6]

These ideas are circumstantially supported by the fact that 'bride-burning', which is relatively common in Northern/Central India where consanguineous marriage is frowned upon, is rare where consanguineous marriage is favoured, as in Southern India and Pakistan.

In the modern world, the level of interest in foetal sex selection reflects the relative value of males and females in society. In Northern India the sex ratio is unbalanced due to selective loss of females (Census 1981). Like others who run genetic prenatal diagnosis services in the UK, we have occasionally been embarrassed by a request for foetal sexing, with the aim of continuing the pregnancy only if the foetus is male. Most of these requests have originated from Indian Hindus; none of the many Muslims attending our prenatal diagnosis service has requested foetal sex selection.

The historical origins and antiquity of the kinship conventions of the Indian sub-continent are unknown (Kapadia 1966), but it does seem that exogamous and endogamous patterns have merged into each other. For example, consanguineous marriage and mutual exchange of sons or daughters are common among Indian tribals and 'scheduled castes' who are not wealthy enough to pay either bride-price or dowry. However, families that become more prosperous often adopt the customs of higher castes, including exogamy with dowry payments (Srinivas 1984). It has been proposed that the population of Northern India was originally exogamous, and that cousin marriage was adopted together with Islam as late as the 17th and 18th centuries. However, since lower castes were particularly attracted by the Islamic concept of human equality, some converts may already have favoured consanguineous marriage.

Cross-Cousin Marriage

Some groups that consider themselves exogamous in fact permit consanguineous marriages because cultural and genetic kinship patterns do not correspond. For instance, Han Chinese consider themselves exogamous and express shock at the idea of consanguineous marriage. However, on enquiry it transpires that this means that people who carry the same name cannot be married. As a woman joins her husband's kindred at marriage and adopts his family name, brothers' children bear the same name, are perceived as relatives and cannot marry each other. However, the children of brother and sister or of two sisters bear different names, are not perceived as relatives and can marry. The Chinese writer Han Suyin describes this common practice in her own 'exogamous' Chinese family (Han Suyin 1968). The same holds true for some Indian castes which forbid marriage if the partners are even very remotely related – but only on the father's side. Conventions permitting cross-cousin but not parallel-cousin marriage embody the concept that the man plants the seed (carrying inheritance), while the woman represents (only) the soil in which it grows.

AFRO-CARIBBEAN KINSHIP PATTERNS

In West Africa, kinship patterns are generally tribal and patrilineal, but Afro-Caribbeans have a predominantly matrilineal kinship pattern. This is because slavery involved systematic and purposeful destruction of the family (Williams 1970). In order to retain the right to sell their slaves independently, slave owners commonly forbade their slaves to marry – and often sold them separately, the children going with the woman. Thus, a situation arose in which grandmothers (or older women) looked after the children while the younger women worked, and men became temporary fixtures moving in and out of matriarchal family units (Foner 1978). Therefore it is not unusual for Afro-Caribbean women to have children by several fathers – easily viewed as promiscuous by Europeans, who often, mistakenly, think a pregnant Afro-Caribbean woman may not know who the father is. In a modern society this pattern can have serious disadvantages for the man, as it deprives him of a central role in the family and can be harmful to his self-esteem. It might also reduce his tolerance of the possibility that he could carry a genetic flaw – which could explain why fewer black fathers than Cypriot or Asian fathers are willing to attend for carrier testing for haemoglobin disorders.

It is conceivable that European families are moving in a similar direction, as women become more economically independent.

PRACTICAL IMPLICATIONS

If we are to reduce misinformation and address the genetic implications of customary consanguineous marriage usefully, we must inform and change attitudes within the 'host' society, and give appropriate information to the groups that favour consanguineous marriage.

For ethnic groups in which women speak little English, medical consultations represent a key interface where women *must* come into contact with the 'host' culture. It is therefore exceptionally important to educate medical and nursing personnel in the roots and nature of cultural prejudice, and to teach them how to handle delicate issues such as cousin marriage.

However, the invasive nature of capitalism exerts immense stresses on the social fabric of societies, including Western societies where family breakdown is considered to be an outstanding social problem. Health workers who criticise traditional kinship patterns reinforce modern society's already severe assault on traditional structures, instead of helping families to cope with medical problems within their own cultural framework and time-frame. The health workers' role is not to adapt societies to medical technology; we should adapt medical technologies to meet the needs of societies.

NOTES

1. As this article presents a personal view, it is written in the first person.
2. Not all carrier combinations involve a genetic risk. For example, if one parent carries alpha thalassaemia and one carries beta thalassaemia there is no genetic risk, while if both carry alpha or both carry beta thalassaemia there is an important genetic risk. Couples of carriers therefore often require expert risk assessment before counselling.
3. Though this is standard practice, it is done more effectively in some areas than in others.
4. In some cultures it extends even to people who are not blood relatives, such as widowed brother- and sister-in-law.
5. In Northern Europe, traces of pre-Anglo-Saxon tribal structures persist in Irish and Scottish clans.
6. Most Europeans are not enthusiastic about such strong family ties, since they restrict individual autonomy. It is true that in a prosperous and stable society this can be a disadvantage, but when society is under threat from war or want, or prejudice, family members tend to cling together. Strong extended family ties often make the difference between life and death for family members.

REFERENCES

Bittles, A.H. 1990 *Consanguineous Marriage: Current global incidence and its relevance to demographic research.* Research Report no 90–186. Population Studies Center: University of Michigan.

Census of India 1981. Series 1. Registrar General and Census Commissioner: India.

Constantinides, P. 1978 The Greek Cypriots: factors in the maintenance of ethnic identity, in Watson, J.L. (ed.) *Between Two Cultures. Migrants and minorities in Britain.* Basil Blackwell: Oxford.

Darr, A. 1991 The social implications of thalassaemia among Muslims of Pakistani origin in England – family experience and service delivery. Ph.D. Thesis, University of London.

Darr, A., & Modell, B. 1988 The frequency of consanguineous marriage among British Pakistanis. *Journal of Medical Genetics* 25: 186–90.

Foner, N. 1978 The Jamaicans: cultural and social change among migrants in Britain, in Watson, J.L. (ed.) *Between Two Cultures. Migrants and minorities in Britain.* Basil Blackwell: Oxford.

Fox, R. 1966 *Kinship and Marriage.* Pelican Books: Harmondsworth.

Han Suyin 1968 *Birdless Summer*, 3rd volume of her autobiography. Jonathan Cape: London. Now available in paperback from Triad Panther books.

Kapadia, K.M. 1966 *Marriage and Family in India*, 3rd edition. Oxford University Press: Oxford.

Modell, B., & Berdoukas, V. 1984 *The Clinical Approach to Thalassaemia.* Grune and Stratton: New York and London.

Modell, B., & Kuliev, A.M. 1992 Social and genetic implications of customary consanguineous marriage among British Pakistanis. *Galton Institute Occasional Papers*, 2nd series, no 4. The Galton Institute: London.

Modell, B., Petrou, M., Layton, M. et al. In press. Audit of prenatal diagnosis for haemoglobin disorders in the United Kingdom: the first 20 years. *British Medical Journal.*

Modell, B., Ward, R.H.T., & Fairweather, D.V.I. 1980 Effect of introducing antenatal diagnosis on the reproductive behaviour of families at risk for thalassaemia major. *British Medical Journal* 2: 737.

Moore, M. 1995 India's consumerism fuels sharp rise in dowry deaths. *Guardian*, 13 April.

Petrou, M., Brugiatelli, M., Ward, R.H.T., & Modell, B. 1992 Factors affecting the uptake of prenatal diagnosis for sickle cell disease. *Journal of Medical Genetics* 29: 820–3.

Report of a Working Party of the Standing Medical Advisory Committee on Sickle Cell, Thalassaemia and other Haemoglobinopathies. 1994 HMSO: London.

Srinivas, M.N. 1984 *Some Reflections on Dowry.* JP Naik Memorial Lecture 1983. Centre for Women's Development Studies. Oxford University Press: Delhi.

Watson, J.L. 1978 Introduction: immigration, ethnicity and class in Britain, in Watson, J.L. (ed.) *Between Two Cultures: Migrants and minorities in Britain.* Basil Blackwell: Oxford.

Williams, E. 1970 *From Columbus to Castro. The history of the Caribbean 1492–1969.* André Deutsch: London.

2 The Work of Culture: An Anthropological Perspective

Marilyn Strathern

In the course of discussing social and genetic implications of consanguineous marriage among British Pakistanis, Bernadette Modell makes a plea for 'counselling that is sensitive to social and cultural background'; she also makes a plea for an 'informed approach' by general practitioners and others (1991: 723). Modell is thinking particularly about the prevalence of marriage between close kin within a wider kinship system which offers a support network. She suggests that the deleterious consequences of close unions should be neither exaggerated nor ignored. What is significant about these remarks is that they envisage counselling as a two-way process of communication. If certain populations need to be informed about genetic risk, medical practitioners and counsellors need to be informed of the circumstances under which reproductive decisions are made. Here she offers a pair of important observations: first, that such decisions are never taken in isolation but belong to a wider nexus of issues; second, that one cannot tell in advance what will or will not be relevant, for that is a matter of investigation. I shall pursue both of them.

The fact that this [Abergavenny] meeting has been called points to the potential of two-way communication. Medical scientists and social scientists may have something to learn from each other; and genetic screening programmes in particular make explicit the need for a mutual flow of information. The flow does not just concern particular populations. If British Pakistanis have 'social and cultural backgrounds', everyone does, and everyone has views that bear on the application of genetic medicine.

I shall start by suggesting where anthropology might have a contribution to make. These days it finds itself in an interesting position. If, as an anthropologist, I endorse Modell's observation that reproductive decisions are not made in isolation, this is partly because, in turn, certain issues that anthropologists had to labour in the past are now taken for granted. They do not have to make a special case for cultural understanding; they do not have to point out the significance of cultural difference or make a discovery out of the diversity of people's habits and lifestyles. It is widely accepted, and especially in the area of family life, that there are important variations in basic assumptions and values, and that people's sense of identity is involved in their being able to pursue what they regard as their own way of doing things. Far from having to convince anyone of the importance of cultural difference, contemporary anthropologists find themselves in a milieu where the very idea of culture signals difference. Indeed, over the last decade culture has acquired a remarkable ubiquity in everyday parlance.[1] However, this very ubiquity poses problems of its own. And here the anthropologist might well wish to be critical of the way culture is used as an explanation for difference.

Culture is a concept of enormous potential in so far as it encourages us to look at what lies behind people's attitudes; but it is not an unproblematic one. Let me ask how the concept of culture is put to work. This paper is divided into two parts. The first outlines what an anthropological perspective on culture might offer; this is followed by a critique which problematises certain understandings of culture. Through the first case I hope to clarify the usefulness of the concept; with respect to the second, by contrast, I shall take to heart Modell's observation that one often cannot tell in advance what is going to be relevant, and make no advance claims about the usefulness of the critique.

AN ANTHROPOLOGICAL PERSPECTIVE ON CULTURE

Culture points to difference, but there is also much more to culture than the recognition of difference. So when anthropologists work with the concept, what work do they make it do? Let me answer by initially noting what is, so to speak, already 'anthropological' in some recent discussions in the field of genetic medicine and counselling.

1. The work of opening up assumptions. What is at issue here is openness to cultural context. It is in this spirit that Modell refers to Aamra Darr's work[2] among British Pakistani (Muslim) families with thalassaemic children where little stigma is attached to inherited disease; the situation appears to be quite different among many families of Indian (Hindu) origin. Seeing these differences as part of people's cultural background adds two dimensions: it reminds one not to take things for granted in this complex field; and it points to the fact that people's sensitivities may be more than a matter of personal opinion when opinions are formed from values or assumptions shared with others.

2. *Working through the ramification of issues.* Such opinions do not exist in a vacuum – they are likely to affect people's dealings with one another. Theresa Marteau's (1991) example of prenatal testing for foetal abnormality is a case in point. The very existence of the tests introduces a shift in the focus of parental concern from an assumption that the foetus is healthy unless there is evidence to the contrary, towards having to prove its health or normality, which in turn places special weight on the knowledge and attitudes of health professionals. What appears to be a solution in one area (preventing 'abnormal' births) may further generate anxieties in another (attitudes towards the disabled) (cf. Price 1990). An anthropologist would see culture constituted precisely in the routes along which such concerns travel. Culture invites us not just to think about the way opinions mobilise values, but to be open to their implications for whole areas of social life.

3. *The work of ideas.* The concept of culture draws attention to the way things are formulated and conceptualised as a matter of practice or technique. People's values are based in their ideas about the world; conversely, ideas shape how people think and react. But there are specific techniques at work here. Anthropologists would argue that ideas always work in the context of other ideas, and contexts form semantic ('cultural') domains that separate ideas as much as they connect them.[3] Angus Clarke (1991) gives us an example of each. Connection: if society places a low value on those with genetic disorders and handicaps, then this idea has a direct effect on the kinds of calculation that health service managers make; here social attitudes afford a context for those calculations, and thus connect values and decisions. Separation: there is by contrast an illuminating divergence between the kinds of evaluations made of parental choice in (say) the context of abortion and those made in the context of sex selection – in the former, parental choice may be primary, whereas in the latter, it may be rendered secondary to other considerations; precisely in the way parental choice is now included, now excluded, as a reference point, we see how evaluations can become compartmentalised between separate contexts. It is worth adding that Clarke suggests a difference between the two sets of contexts: 'western cultural values' prevail in the first whereas 'the social preferences of other cultures' are ignored

in the second. Culture itself thus works as an idea: when we introduce it into our explanations it shapes the perception of cause and effect in a specific way; we may also find it being used to different effect in different contexts.

4. The invitation to work with specifics. A culture is always specific to time and place. Anthropologists are interested in how views of the world belong to the time and place where people live, and thus operate relative to their particular circumstances. This would suggest that anthropologists are not interested in universals. Of course they are: it is just that no one leads a universal life – whatever we share in common with the rest of the world, each of us lives it out specifically in the here and now. This approach should have much in sympathy with the investigation of genetic conditions as medical disorders, for disorders only become apparent under specific environmental and aetiological circumstances, even if it is only the age of the sufferer at issue. In multifactorial diseases, place of residence may literally be significant. Sarah Bundey (1993) has noted an increase in the profile of diabetes mellitus and multiple sclerosis among the children of people from India and the West Indies: the difference is that the second generation have grown up in Britain.

If anthropologists are interested in the details that make all the difference to people's lives, they have developed the study of kinship as an area which demands attention to specifics. They see significance, for example, in what people call their close kin – whether a father's brother is equated with the father or whether a father's brother and mother's brother belong, as they do in English but very often not elsewhere, to the same category. These are not trivial differences: they are clues to the ways in which people conceptualise relations between close kin. The same is true of ideas about conception. Indeed, anthropologists are familiar with the idea that there are cultures where a man may be regarded as an agent in a child's birth and yet not share substance with it, a birth father who is not consanguine. Many years ago now, and long before the advent of the new reproductive technologies, Edmund Leach (1961), in commenting on such a state of affairs for certain Oceanic societies, asked whether there might be any culture where the birth mother might be so regarded. He found a case from Asiatic ethnography: whereas the father is counted as a relative by substance in this case, the mother who gives birth to the child is counted as its relative by virtue of her marriage to the child's father. It is in the specifics of the connections and separations people make that anthropologists recognise culture.

5. As an aid to reflexivity. In undertaking the study of cultures outside the Euro-American orbit as well as within it, anthropologists become attuned to the assumptions with which they themselves start. Indeed they are frequently led to specify what of their own circumstances makes them cultural beings. So, in the account just given, I have used the terms 'mother' and 'father' as an English-speaker would, to denote the procreators of a child, even though I have used them for situations where procreation is not the most significant aspect of parenthood. The non-consanguineous birth father and birth mother are what we could call parents to the child – they have obligations to it, nurture it and generally 'parent' it – but that parenthood is not felt to rest on a tie of substance; in both cases, it rests instead on a bond much better translated as 'affinity'.[4] Yet it is clear that all these translations are approximations. Whether we call someone 'father' or 'male parent' or 'affine', none of the English terms offers a complete rendering of the vernacular. This, in turn, makes anthropologists aware of the connotations of their own vernacular. If in English parlance the mother's consort (husband/sexual partner) is presumed to be consanguineous with the child, that is a specific presumption.

Finally, and as an outcome of all these points, anthropologists would argue that there is *no substitute for empirical enquiry.* Whether or not they call it 'fieldwork', they would insist on the importance of investigation. And that is because they are also open to the idea that you often cannot tell in advance just how things will work out. This is particularly true in an area that touches on kinship and family life: information about one ethnic community will not substitute for information about another! At the same time, they do not go so far as to say that the same could be said of every individual. Anthropologists are interested in those aspects of people's circumstances that belong to the relationships between them; culture comprises the techniques and practices which mediate people's interactions and communications with one another. Culture, in this view, acts as both a resource and a limitation, both as the enabling technology through which communication takes place at all and as a constraint on what can be said or done. A person who classifies a father's brother in a different way from the mother's brother may also treat the two persons in quite different ways and perhaps reckon the one a closer kinsperson than the other. But whether they do or do not has to be found out and cannot be deduced from the classification itself. Only empirical enquiry will do.

No doubt this is already obvious. Anyone drawing a genealogy has to be alert to issues of classification. He or she will need to know the cultural procedures of categorising kin, as he or she will need to find out the role that procreation plays in the conceptualisation of relatedness in order to translate what a person means when they say so-and-so is the child of so-and-so. Within this framework, he or she will also need to know how knowledge is distributed between people, and the different interests that different parties may have in that knowledge. Above all, anyone detailing family history in order to uncover the relationships between individual members will want specific, not generalised, statements (cf. Richards 1993). One would like to think that the 'anyone' could be equally a social scientist or a medical geneticist.

Anthropology's contribution lies in its being able to bring together, under a model of how culture works, insights that practitioners may already have or methods they already employ; it does so in a way that allows one to specify the interrelationships, complexities and contradictions in the elements that make up people's 'backgrounds'. There is indeed much more to culture than simply the recognition of cultural difference.

A CRITIQUE OF THE CONCEPT OF CULTURE

Now the anthropologist, conscious about the way he or she uses concepts, is also interested in the way concepts behave at large. If anthropologists operationalise the concept of 'culture', they are certainly not alone in this and have no monopoly over its usage. That need not stop them from being critical. It would be right, for instance, to be wary of the extent to which 'culture' has become a synonym for 'difference'.

I have been stressing the kinds of connections that could be made between anthropological procedures and issues raised in the literature on genetic medicine. From a quite separate vantage point, I shall now sketch in some general observations that lead to just two points. The two points are two blind spots in some of the current debates.

One important criticism is by no means confined to anthropology, namely the way *difference may be used as a gloss for discrimination*, and *culture* as a gloss for *social* conditions. 'Multiculturalism' can obscure issues of race (Stolcke 1995), much as the idea of 'working-class culture' can conceal conditions of social deprivation. The expectation is that people from particular backgrounds will be locked into certain ways of doing things because of their attachment to that background – rather than (say) because of the social conditions they find themselves in. The very concept of 'minority culture' contributes to this expectation. In that context we encounter a tendency *to ascribe whole persons to their culture.* This is most true when the culture in question seems to belong to an ethnic group: it happens, for instance, when ethnicity is taken to be a person's defining characteristic. However, there are many situations, by contrast, where the concept of culture is used instead to refer to *very partial aspects of people's lives.* If one talks of corporate culture, for instance, or the culture of the classroom, one does not mean that people who work for ICI or who go to school are only that: rather the connotation is that the particular location has its own rules and habits, so such rules and habits define the location rather than the person. Now this difference in usage is not neutral. The contrast between these two very different appeals to culture endorses a fundamental divide. I state the divide crudely, but there is no other way of doing so: the divide so created is between ethnic 'minorities' – where whole persons are defined by their culture – and 'the majority' – who see themselves instead moving between different cultural locations. An outcome of this is that it sometimes seems that minority persons appear to have 'a' culture while the majority do not. Hence the perpetuation of the majority view that what is significant about minorities is their difference.

So what are the two blind spots? The first relates to the construction of understandings about scientific matters: *culture may be seen to get in the way of scientific understanding.* That is, a person's failure to endorse something or be open to suggestion may be put down to a cultural block, especially if that person can be ascribed to an ethnic group. The field of reproductive medicine provides an example in problems about the shortage of 'ethnic' eggs. Yet if members of an overseas Chinese community explain that they have feelings about ancestry which would be violated by egg donation, to treat these as cultural beliefs or 'customs' may be to treat them as less than reasons. For it is 'culture' that instead appears as the reason why there are so few 'Chinese eggs' available for donation.[5]

The problem here is the implication that culture is a handicap that needs to be overcome: if these people changed their customs ('Westernised' more), they would not put such barriers in the way of medical need. Such a construction devalues the way in which people's views of the world are indeed informed by a variety of suppositions which the anthropologist would call cultural – you cannot just cut them out and dispense with them as the notion of custom suggests. At the same time it ignores the fact that any such discourse is mutually constructed (see Parsons & Atkinson 1993). You are not listening to the voice of Chinese culture as such: you are listening to people's dealings with medical knowledge and clinical practice. People live in a media world, exposed to information of all kinds, including information about developments in medical science, and such a response to the problem of egg shortage is formulated in that interactive context. Even when the response takes the form of an appeal to 'custom', custom may not be the reason for their objections. Someone's appeal to Chinese ancestry may, in fact, be as rhetorical as another person's drawing on the legitimacy of individual expression in claiming to be squeamish on 'personal' grounds. For whatever else is happening, the views are also in part constructed by the scientific discourse itself, and people's reasons do not necessarily imply an opposition to scientific knowledge as such. To suppose an opposition exists is to suppose that culture is sealed off from scientific knowledge; this leads to the second point.

The second blind spot is a converse of the first. If too much is made of minority cultures then *too little may be made of the majority culture*. If in the first case culture weighs people down with 'customs', in the second it may be trivialised as 'lifestyle'.[6] Like corporate image, lifestyle is difference by choice. Again we encounter the notion that if you can identify a culture you can also change it, only here difference has become a matter of consumer preference. This view of culture again devalues the embeddedness of people's ideas and the implicit nature of the assumptions they bring to debate. Yet it seems to me that there is a particular onus on those who speak from the vantage point of the majority culture to make their own cultural foundations explicit. An example comes from the Nuffield Report on ethical issues in genetic screening, which is worth examining at length; neither culture as custom nor culture as lifestyle help the elucidation, but culture understood as those contextualising practices outlined in the first part of this paper provides considerable critical purchase.

The Report of the Nuffield Council on Bioethics notes the complex range of ethical issues that need to be exposed to public debate. Chapter 4 deals with the meaning and implications of informed consent in the context of screening. The Report admirably points out that, by contrast with many medical procedures, such screening introduces the family as the object of enquiry. The screening is, after all, about how a person is (genetically) related to other members of the family;

reproductive decisions never concern an individual alone. An anthropologist would endorse this wholeheartedly.

The Report then defines 'family' in para 4:2:

> 'Family' needs to be understood as covering an extended set of
> relatives linked by blood ties or by marriage or by both. Members
> of families may or may not be in close touch. They may live far
> apart... and may sometimes be unaware of the relationship. (1993:
> 29)

The assumption is that such persons are related whether they know it
or not. If we talk of relatives by marriage, this might mean that distant
cousins make marriages one never gets to hear of. If we talk of persons
linked by blood ties, this must mean that the ties relate them regardless
of their social relationships. So 'relating' in this sense does not mean
recognising and acting on a tie but being tied in a non-social manner.
In the context of a document on genetic screening the nature of that tie
is further assumed – it becomes a genetic connection between persons
who share one another's hereditary ancestry though they may have no
knowledge of it. Indeed, the paragraph continues in just this vein:

> Nevertheless they [members of families] may share important
> genetic traits. Genetic screening may [as a consequence] discover
> information about persons who have neither been screened nor
> consented to be screened [i.e. persons otherwise out of touch].

The problem of imperfect knowledge appears to be compounded by a
problem that stems from people's lifestyles (not all family members
may be contactable). For where ethical practice rests on informed
consent, it becomes troublesome to uncover information about gene
carriers who are genetically close but who, because of the vagaries of
family history, are socially distant.[7] The genetic closeness may, of
course, be as close as that of an otherwise unknown father.

There is more at issue than the awkwardness of family relationships.
What is this term 'blood tie'? *Blood* tie? No doubt the Report wishes to
make its findings accessible by drawing on a homely and familiar
idiom. But to equate blood tie with genetic tie requires cultural knowledge. That particular connection – between consanguinity and biology
– belongs to a particular culture, one I would call Euro-American, and
the work it does in that culture is ideological. Any cultural precept
can, of course, have ideological status – can be pressed into the service

of persuading people to a particular course of action or way of thinking. In the present case I mean that one kind of relationship (social ties
between family members) is justified by its grounding in another
(natural ties of biological connection). Leaving marriage and adoption
aside, family members are thought of as related 'by blood'. This piece
of ideology supposes that common biological substance both does and
should flow in the arteries and veins of relatives.[8]

I do not wish to add complexity to what is already perceived as a
most complex ethical area in relational terms – that is, the consequences for others of screening individuals. I merely wish to illustrate
the workings of ideology. We might remark, for instance, on how the
prior (natural) status of biological links is used to justify presumption
of common interest. Chapter 5 of the Report uses the terms 'family
members' and 'blood relatives' more or less interchangeably, and
imagines them forming a group: that is, acting together as people
ordinarily (normatively) in contact with one another. Such closeness
is further taken to imply a bond of mutual interest (though it is
acknowledged that 'the perceived interests of members of the same
family sometimes clash' [5:5]). So although, in practice, malice between
relatives is as much part of kinship behaviour as its opposite, the
Report envisages circumstances where the individual family 'member'
must sacrifice his or her right to privacy for the sake of the group
(5:30). The demands of genetic knowledge thus prioritise the fact of
shared substance over how persons conduct their relationships.

Ideology works to conceal as well as reveal the conditions of people's existence, and its diffuseness is accommodating. The idiom of
'blood relative' is happily unspecific. It does not follow, of course, that
family members necessarily have genes in common in the way their
social relationship might suggest. Yet all the comments in the Report
on confidentiality and information among family members rests on the
assumption that blood kin are what they say they are. The question of
non-paternity is raised only once (5:31) and the question of anonymous gamete donation not at all.

The idiom of 'blood tie' offers a particular cultural model of and for
relationships. It contains two overlapping but not identical proposi-
tions. On the one hand, social parents are presumed to be the procreative parents unless shown otherwise; on the other hand, the act of
procreation in itself makes 'parents' out of the providers of hereditary
material. The consequent dilemma, then, to which the Report refers in
the paragraph that mentions non-paternity (5:31), is a *cultural*
dilemma.[9] The reference is to the burden of responsibility on those

in a position to disclose genetic information. If information might yield compromising evidence about paternity, the dilemma is precisely that of a culture which assumes a father and child share a genetic bond. Disclosure of non-paternity would take away the child's 'father', while failure to disclose would withhold information from the presumed father on the basis of which he might have made further reproductive decisions. The problem arises because of the initial cultural supposition that family members are genetically related and that this is a factor in their reproductive decision-making.[10] *Culture suddenly appears to be an impediment to clarity of information about genetic connection; yet if it does so, it also furnishes the very inspiration for a popular interest in genetics in the first place.*

If the dilemma is born of a culture that presumes a family is made of persons related through procreation, then the same culture also fuels the interest would-be parents have in striving to have children of their own. They want the parent–child relationship to be of a particular kind. Where that relationship is presumed to rest on biological connection, persons may turn to reproductive medicine for assistance, and developments in this field have, in turn, stimulated research into genetic therapy.[11] It is a genetic model of relationships that has in part, then, led to the development of such techniques, and that genetic model is part of the kinship system of a particular (Euro-American) culture, which scientific interventions then appear to endorse. But the very acceptability of such interventions, the cultural endorsement they receive, in turn, from these Euro-American models, will, in the long run, be a challenge to the interventions themselves. It will test people's assumptions about the place of biology in the creation of relationships. What 'new kinship' will result, we do not know, but the prospect should be intriguing for genetic medicine.

* * *

To conclude: (1) The field of debate stimulated by the ethics of genetic screening holds great promise in making us aware of relationships – both between persons and between the ideas and values that anthropologists would claim for their concept of culture. Indeed, it must be unique in the medical field in the sophistication of the social and cultural commentary it has attracted. (2) I would firmly add that we should draw no line between those situations where we see the need for cultural sensitivity and those where we do not. (3) At the same time, sensitivity must include a capacity for critique: it is no good just promoting the concept of culture because of its virtues in portraying the value of pursuing relationships; one has to keep in mind that it can also be used to exclusionary effect. We need to be alert to those situations where a selective interest in cultural difference actually makes it harder to take what people voice as a matter of judgement and reason. We also need to be alert to unspoken cultural assumptions of the kind that underlie the way ethical debate proceeds. (4) Finally, there is no single road here, and I finish with a plea for openness that includes openness to what cannot be anticipated. It is offered in the spirit of two-way communication.

It is not always possible to anticipate what information is going to be useful. Certainly cultural analysis cannot stop at what seems immediately to be useful, and should not so stop, for what is defined as useful is often part of the problem. We need to be open to this. But we also need to be open to current practice. I have not held back on my criticism – including criticism of the way the very concept of culture itself is sometimes used and of the way it is sometimes ignored. Like any piece of technology, the issue is how we decide to make the concept work.

NOTES

This is the text of the spoken contribution to the Abergavenny conference and is offered in the spirit of debate.

1. It was a theme running through the 4th Decennial Conference of the Association of Social Anthropologists of the Commonwealth in 1993 (Oxford), and was the subject of a panel at the 1994 meetings of the Association of European Anthropologists (Oslo). For an important statement see Stolcke (1995).
2. Modell is referring to Darr's contribution to the CIBA conference and, in the following example, to her own report on certain Hindu families. See Chapters 1 and 5 in this volume.
3. The way in which certain kinds of knowledge are connected to or separated from one another in debates over the 'new reproductive technologies' is discussed at length in the contributions to Edwards et al. 1993.
4. Indeed, the affinal parent's nurture may be thought of as a gift to the child's consanguineous kin which must be repaid with gifts in due course.
5. I have fictionalised the case which follows. Its inspiration comes from Frances Price's commentary on a BBC Radio 4 programme which 'included as news... the pressing need for more sperm and egg donors to help infertile couples to have children. The *Today* item highlighted the shortage of donors from black and Asian people, and attributed this apparent unwillingness to donate to a distinctive, non-normative concern with questions of 'ancestry' or family relationships – to "cultural" differ-

52 Culture, Kinship and Genes

ence.' She goes on to ask what might be at stake when ideas about relationships and generation are seen to stand in the way of medical intervention in the making of families. I am grateful for permission to quote from the unpublished paper 'Seeing things differently: representational practices with the clinic' by Frances Price (University of Cambridge, 1994).

6. There are resonances here with the kinds of understandings against which Malinowski (1922) saw himself doing battle – in his case the view of 'savages' as either oppressed by culture ('custom') or having no culture at all.

7. I use 'troublesome' in the double sense of being troubling on the several ethical grounds which the Report considers and causing trouble for those who work to give the best counselling possible.

8. Blood as an idiom for connections of bodily substance is widespread. This includes Europe: see S. Franklin and M. Strathern, Kinship and the New Genetic Technologies: An assessment of existing anthropological research, Report prepared for the EC Human Genome Analysis Programme (1993).

9. The dilemma described in the Report is between giving a man access to information that will affect his reproductive decisions in the future, and protecting the woman who might be harmed by the inevitable questions about her other relationships.

10. One does not want to exaggerate the phenomenon. However, Modell (1990: 478) assumes that non-paternity is a likely finding in screening programmes. What is culturally illuminating is that is should be an issue at all.

11. The consultation document processed by the Human Fertilisation and Embryology Authority on Donated Ovarian Tissue in Embryo Research and Assisted Conception (1994) notes that, under the HFE Act, donated eggs can be used to create embryos for research, including increasing knowledge about the cause of congenital disease and developing methods for detecting gene or chromosomal abnormalities in embryos before implantation.

REFERENCES

Bundey, S. 1993 Genetic diseases in UK ethnic minorities. Paper delivered to the British Association, Keele.
Clarke, A. 1991 Is non-directive genetic counselling possible? Lancet 338: 998–1001.
Edwards, J. Franklin, S., Hirsch, E., Price, F., & Strathern, M. 1993 Technologies of Procreation: Kinship in the age of assisted conception. Manchester University Press: Manchester.
Leach, E. 1961 Rethinking Anthropology. The Athlone Press: London.
Malinowski, B. 1922 Argonauts of the Western Pacific. Routledge: London.
Marteau, T. 1991 Psychological implications of prenatal diagnosis, in Drife, J.O., & Donnai, D. (eds) Antenatal Diagnosis of Fetal Abnormality. Springer-Verlag: London.

Modell, B. 1990 Cystic fibrosis screening and community genetics. *Journal of Medical Genetics* 27: 479–479.

Modell, B. 1991 Social and genetic implications of customary consanguineous marriage among British Pakistanis. Report of a meeting held at the CIBA Foundation on 15 January 1991. *Journal of Medical Genetics* 28: 720–3.

Nuffield Council on Bioethics 1993 *Genetic Screening: Ethical issues.* Nuffield Council: London.

Parsons, E., & Atkinson, P. 1993 Genetic risk and reproduction. *Sociological Review* 41: 679–706.

Price, F.V. 1990 The management of uncertainty in obstetric practice: ultrasonography, *in vitro* fertilisation and embryo transfer, in McNeil, M. et al. (eds) *The New Reproductive Technologies.* Macmillan: London.

Richards, M. 1993 The new genetics: some issues for social scientists. *Sociology of Health and Illness* 15: 567–85.

Stolcke, V. 1995 Talking culture: new boundaries, new rhetorics of exclusion in Europe. *Current Anthropology* 36: 1–24.

3 What is an Ethnic Group? A Biological Perspective

Helen Macbeth

The first point to highlight is the risk that the words 'ethnic group' may have been used too simply to replace the word 'race', when concepts of race are considered to be both politically unacceptable and biologically inaccurate. It is not the objective of this paper to discuss political acceptability, but the same difficulties with the idea of discrete biological populations that invalidate traditional concepts of 'race' are central to the discussion of ethnic groups and genetics.

A discussion of the word 'population' in the species *Homo sapiens* provides a useful introduction to a biological review of the question 'what is an ethnic group?'. Whatever ethnicity is, it is related to a perception of groups of people as discrete entities, some would say 'populations'. This is a topic which requires a *biosocial* approach. However, the history of attempts to unite those educated in biological sciences with those concerned with social perspectives in order to achieve *holistic* understanding has been full of misunderstanding. One cause of problems has been the ambition to produce a neat, integrative biosocial science, which is almost certainly impossible as no neat, integrative philosophies unite the social sciences in the way that the ideals of 'scientific proof' and the body of biological knowledge based on such 'proof' unite biology. The point here is that lack of one integrative biosocial science is a red herring and should not hinder biosocial cooperation that produces cross-disciplinary perspectives on the human condition. This will be discussed further in the last part of this chapter where, despite the problems, some optimism lurks.

POPULATIONS

Anthropology, including biological, social and biosocial strands, is a discipline concerned with human populations rather than individuals, but in all but the most isolated cases the definition of a population is problematic. It is ironic that this discipline is centrally concerned with something that it fails to define adequately. In 1976 Gomila exhaustively discussed the meaning of the word 'population', including biological and social concepts, while at much the same time Ardener (1974) tackled the same problem from the social anthropologist's point of view. Both concluded that, except in rare cases of isolation, the population could not be defined. Maybe that is why traditional anthropologists, both biological and social, sought isolates in remote regions. Two papers in these proceedings address the question 'what is an ethnic group?'. That question incurs the same problems as 'what is a population?'.

Many readers of this book will have a training in biology, and particularly in genetics, which will make the words 'Mendelian population' familiar. For others, Dobzhansky's (1962) definition of a Mendelian population – as a group within which the majority of individuals find mates for reproduction and the procreation of the next generation – will provide a useful summary. Biologists will be familiar with the increasing divergence in gene frequencies[1] that occurs on either side of some barrier to passage and to reproduction. It is, therefore, common to associate the limits of populations with the boundaries that cause barriers to passage and so to mating, for example mountain ranges, seas, rivers, etc. Yet distance itself is a barrier, as Wright showed in his (1943) model of genetic isolation by distance.

Humans have been able to cross every physical obstacle on this planet as well as great distances. Nevertheless geographic features, such as seas or mountains (e.g. Garcia-Moro et al. 1986), or distance (e.g. Boyce et al. 1968), can be shown to reduce marriage frequency. One might postulate that for many species mating frequency may be correlated to energetics: that is, where the energy needed to cross or bypass a physical obstacle equals the energy of crossing a greater distance without obstacles, mating frequencies would be similar. However, in the case of humans there are problems with this simple equation. An important characteristic of the species *Homo sapiens* is the rich complexity of our social structures, with social bonds and social barriers between individuals and between groups of individuals. Humans not only mate, they marry. Without here trying to define marriage, it can be said that in all human societies choice of marital partner is a serious social issue. Furthermore, most procreation of the next generation occurs within some form of marital arrangement. In this way, socially perceived obstacles to marriage and social distance can reduce gene flow in a similar fashion to physical barriers and distance. Marriage rules are social rules. Being socialised into believing themselves to be part of a given group the majority will tend to marry according to the rules of that group and pass on their genes within that group. This sounds very 'Mendelian'. Yet, please note that in this description it is the biological variable that is the dependent variable, and socialisation the independent variable: social bonds, boundaries and barriers affect marital choices, therefore mating, therefore gene frequencies.

SOCIAL GROUPS

While genetic models, constructed during study of non-human species and barriers, throw some light on human population genetics, they are quite insufficient because human passage across such barriers depends not only on the energetics and mechanics of crossing, but also on social attitudes to the barriers they represent. Delineation of human populations by physical barriers or by cartographic lines may be a useful classificatory tool, but it may have no Mendelian[2] significance.

If dependent upon social structures, is the human biological population, then, just the social group? Try to define the social group and further problems are found. Depending on the complexity of any society, there are likely to be family, lineage, clan, caste or class subgroups within the society for individuals to identify with, as well as adjacent groups sharing perhaps language and divinities, which allow a still wider classification. Because of different social links, the extent of one person's social group may not be the same as their neighbour's. These problems are related to concepts of identity. The author of this chapter is Scottish in England, English in Scotland and a 'pom' in Australia, as well as European, and that just describes identities due to regional affiliations. Support of Kirtlington village can be contrasted to Bletchington in local football matches, or of Oxford University versus Cambridge, or of Oxford Brookes, up the hill, rather than that other university in Oxford. Frequently, identity is signalled by what one is not, but this may be context dependent.

The point is that everyone has these different aspects of their identity. Subgroupings, as well as larger clustering of identifiable groups, make a 'society' as difficult to define as a 'population', and sociologists (e.g. Thomas 1938) have similarly struggled with defining a 'community'. So, if one argues that the mechanisms that define the boundaries of a human Mendelian population are social, then one is only faced with the problems of defining the social group, which should be left to the social anthropologist (see Sharma in Chapter 4).

As Chapman (1993) pointed out, every social anthropologist would have something to say about ethnicity – but they may well not agree. Even reviewing its derivation from the ancient Greek *ethnos* does not help greatly. Homer used it for large undifferentiated throngs, whether non-human or human, but Chapman discusses its progress in classical Greek from Homer's use to the later use, meaning foreign or other tribes, or even those provinces viewed as less 'civilised'. This relates understandably to its use in New Testament Greek where it refers to non-Jewish, non-Christian people and became translated as *gentilis* in Latin (gentile). Yet, in modern Greek, *ethnos* is a word which is used to describe the Greeks themselves possibly because it evolved from a time when the Greeks proudly viewed themselves as a minority that was significantly 'other' within the larger Ottoman Empire. In this light, the links with today's discussions of minority groups and their rights seem appropriate.

ETHNIC GROUPS

This introduction to the concepts of 'a human group' and 'ethnicity' has been crudely abbreviated but was a necessary preamble to this discussion of what an ethnic group is. One of the biggest problems with use of the words 'ethnic group' today is that in different contexts the meaning may be different and yet not understood to be different. As mentioned above, the phrase is commonly used to replace the word 'race', which probably derives from a social sensitivity about racism. In anglophone society a large majority of people would not wish to be called racist, but still carry an entrenched view of the human species divided into discrete racial groups. It should not be forgotten that such a concept is equally fundamental to, and important for, political pressure groups fostering the advancement of, say, 'blacks', but the idea of discrete races cannot be defended biologically because of the clinal distributions of gene frequencies.[3] Meanwhile, we, the biologists, have hardly been evangelical in spreading the word about *clines* in gene frequencies. The effect of such clinal frequencies will be discussed below. To others, the words 'ethnic group' are not a euphemism for race, but a conscious attempt at group classification on the basis of culture and concepts of identity.

Three components of ethnicity can be distinguished in order
to suggest how its social complexity could be summarised for cross-
disciplinary communication (Macbeth 1993). Ethnicity is found:

1. 'in the processes and transmission of ideas of belonging to a group
   by those IN that group';
2. 'in the way people OUTSIDE that group classify people into that
   group'; plus
3. 'in all the features and factors which confirm, and which are recognised as confirming, the group classification both to those inside
   and to those outside that group'.   (Macbeth 1993: 47)

Various classifications of who is an Australian Aborigine exemplify
these processes. The official Australian government definition now
refers to those individuals recognised as being of Aboriginal descent
by the person classified and by other Aboriginal people. It is the self-
recognition of the identity that seems important here, and yet the
words 'of Aboriginal descent' imply that that recognition is based on
a concept of biological ancestry, even when the individual's visible
facial features do not correlate very highly with the outsider's stereotype of an Australian Aborigine's 'typical' characteristics. Only the
most cursory glance at the faces of those identifying themselves as
'Aboriginal' groups, demonstrating against the 1988 'bicentenary cel-
ebrations', confirmed the diversity of facial features, skin colours and
probable ancestry.

Two discussions arise from the way this last example was worded.
First, the Australian official definition appears at first glance to fall
within (1) above about self-identification, but it is a self-definition
within the terms of the European-based culture of official Australian
policy. Yet, Aboriginal within-group classification more commonly
refers to dreamlines, languages, and the real and mythological ancestry
of families. The reader will note the concept of ancestry even here.
Meanwhile, the most common outsider's classification is still based on
a stereotype of the skin colour and facial characteristics of an Aborigine. The first point about the outsider's classification, both official
and stereotypic, is that it ignores the concepts of families, mythological
ancestry and dreamlines. The second point is that it has to stereotype
to ignore the diversity that is actually there.

It is typical of outsiders of any group not to perceive subdivisions
recognised as significant within the group. How aware are most resid-
ents of Europe of the diversity of language, religion and self-identities

within Africa or the Indian sub-continent – or of how that diversity becomes translated in people who now live in one of the European nation states?

The lack of perfect congruence between the group definitions by those inside and those outside 'ethnic groups' provides one explanation of why the groups cannot be delineated. This lack of delineation has genetic consequences, discussed below.

The second area of discussion arising from the Australian example is the common use by outsiders of biological features for their social classification. This appears to bring the biological back as the independent variable and the social as the dependent variable, because someone with certain biological features would be classified as an Australian Aborigine. However, human biological diversity is infinite; genetic variation *within* groups, however defined, was shown by Harris's work on polymorphic proteins (1980) to be greater than variation *between* populations. In the face of such diversity, we stereotype in order to classify people, but stereotyping is a mental (and cultural) process, noting some features, ignoring others. If what people are conscious of seeing amidst the continuous diversity differs, the delineation of the group is likely to be variable. So, stereotyping may only simplify the definition for each individual observer, for it is not objective.

Self-identification remains the preferred basis for assignment to ethnic groups, and as it has the strongest effect on marital patterns and so gene frequencies, it is of interest to the population geneticist. Even so, the effect on marriage and other mating will not be one of total control. Thus, ethnic groups do not become sealed containers of idiosyncratic genes. Clinal distributions of gene frequencies may become a bit steeper at language boundaries (Barbujani & Sokal 1990) but there is still some gene flow. Part of that gene flow is due to individuals who consciously marry out of their social group, but part is because of different perceptions of the ethnic boundaries.

There may be an applied medical reason for interest in the biological outcome of these perceptions of ethnicity. Take, for example, an interest in Tay-Sachs disease. Many recognise that this genetic condition is more common in Jews, but who are the Jews? Do they constitute an ethnic group? Maybe a race? Or should they just be considered a social group or adherents of a given religion? Must one inherit Jewishness to be a Jew? One can convert to the religion of Judaism – does that make one a Jew, or one's children Jews? To some, the classification of one's mother is highly important, but on what basis should she be classified? Among Orthodox Jews there is a clear conceptual opposition between 'the chosen people' and the rest, but is this reflected in their biology?

Jewish identity, throughout the Diaspora worldwide, has been ritually reinforced over centuries. Yet now reunited within Israel, the social, cultural and religious (Paine 1989) diversity has become clear. Israeli medical literature is full of so-called 'ethnic' differences in health conditions within Israel (e.g. Dolberg et al. 1991, Kalter-Leibovici et al. 1991, Rozen et al. 1987, Stein et al. 1988). Studies of gene frequencies have shown the genetic differences between Jewish groups around the world (e.g. Chakraborty & Weiss 1982, Morton et al. 1982, Mourant et al. 1978) and, to return to Tay-Sachs disease, it is found in clustered frequencies only among the Ashkenazi Jews of European ancestry. Also, all Jewish groups studied showed the genetic effects of miscegenation with their geographic neighbours. Such a situation is expected in the cases of all other 'ethnic groups'. Endogamy is unlikely to have been as total as some people might assume, or claim, and this leads to gene flow and so clinal variation in gene frequencies.

There is an aspect of clinal patterns of which the geneticist should be aware. Once boundaries have been drawn, on whatever basis, in order to identify a division of groups within a clinal pattern of change in gene frequencies, significant genetic differences between groups are likely to be shown. Figure 3.1 demonstrates this point: alternative 'boundary' lines, A and B, are drawn vertically through a clinal pattern of change in proportions of 'x' and '–'. Either line would result in a significantly different combination of the symbols between the two groups.

So the evidence of group differences does not necessarily confirm the biological significance of any boundary lines chosen, because in a clinal pattern almost any other boundary would also have shown significant genetic diversity between sectors either side of that boundary. This is a very important social point too. In real situations, care should be taken in explanation lest the genetic results be used socially to report some biological validity of the group delineation first chosen. Perceptions of ancestry are frequently used as the rationale of social classification into ethnic groups. So any suggestion of difference in gene frequencies may be used to support claims that the group is defined by its discrete and idiosyncratic genetic characteristics and the genetic diversity within the group could be ignored.

This topic cannot be left here, giving the impression that concepts of ethnicity have no effect on gene frequencies. Social affiliation does indeed lead to gene clustering; it is only the boundaries that cannot

A   B

```
- - - - - - - - - - -|- - - - - - - - - - - -|- - - - - - XXX
- - - - - - - - - -|- - - - - - - - - - - -|- - XXXXXXX
- - - - - - - - - -|- - - - - - - - - - - -|XXXXXXXXX
- - - - - - - - - -|- - - - - - - - -X|XXXXXXXXXX
· - - - - - - - - - -|- - - - - - XXX|XXXXXXXXXX
- - - - - - - - -|- - -XXXXXXX|XXXXXXXXXX
- - - - - - - - -|XXXXXXXXXX|XXXXXXXXXX
- - - - - - - -X|XXXXXXXXXXX|XXXXXXXXXX
- - - - - -XXX|XXXXXXXXXXXX|XXXXXXXXXX
- - - XXXXXX|XXXXXXXXXXXX|XXXXXXXXXX
```

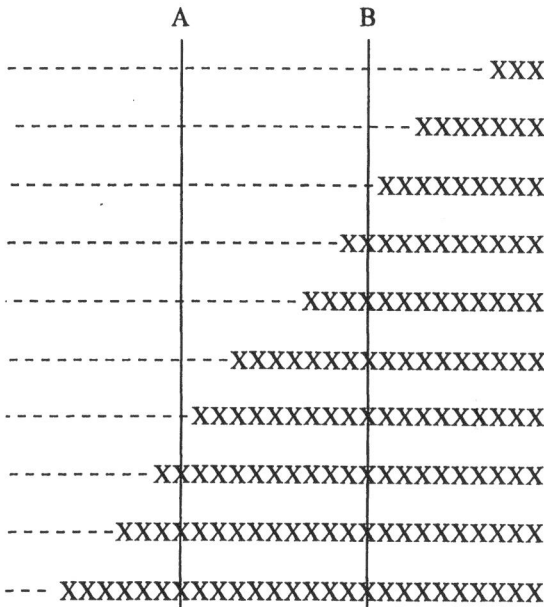

Figure 3.1   In this figure, there is a gradient across the page in the relative frequencies of two symbols, – and x. If two boundaries are set (at A and B) to divide the figure into three groups (populations) then there will be marked differences in the frequencies of the two symbols between the three populations. The boundaries, however, and hence the groups, are entirely arbitrary. There is a continuous gradient (a cline) in frequency of the symbols across the figure. In genetics, we may talk about a gradient in gene frequencies across a continent. Group differences will be found wherever a boundary divides a cline.

be defined. Where the self-identification of an ethnic group is based on culture and ancestors from a distant part of the globe, some genes that cluster in that group are very likely to be different from those in the recipient population. This is due to the processes of biological adaptation of the immigrants' ancestral population to the environmental characteristics of that distant region. Then, migrants who can find enough others of similar background will tend to keep together socially and retain language and other cultural practices. This would include, to some extent, preferred marriage patterns within the group, thereby affecting gene clustering.

Finally, it should be remembered that the clustering of biological characteristics may be greater than the clustering of genes, because multifactorial characteristics are also affected by lifestyle. The likely congruence of lifestyle, cultural practices and self-identified ethnicity could cause more homogeneity in the multifactorial characteristics of the ethnic group than in the genes, because of the inevitable genetic diversity within any group. Despite their probable knowledge of genoclines,[4] human biologists are themselves members of society and socialised into discussing the socially defined ethnic groups as units. Once described as a unit, means and variances of biological variables in any ethnic group can be compared with, and shown to be statistically different from, those in any other group or in the majority population. Many of these characteristics will be multifactorial and a few may have no genetic component at all, and yet, unless worded carefully, the human biologist may add to the concept that the ethnic group is not only biologically different, but a genetically discrete population. Such an inaccuracy is very similar to the popular and incorrect view of 'races'. So while knowledge from human biology provides the strongest evidence against biologically discrete 'races' in the clinal distribution of gene frequencies across land masses, the biologists themselves may increase the misconception by focusing on the statistically significant differences in biological characteristics, without explaining multifactorial aetiology and without questioning how they have chosen to define the populations compared.

MEDICAL UNDERSTANDINGS OF 'RACE'

Despite the potential for confusion, understanding those statistically significant biological differences may be of medical importance and should not be ignored. Using only crude classifications of 'ethnic groups', epidemiological differences are found and the medical literature on these differences has increased greatly since the early 1980s. While the aetiology of differences ranges from genetic clustering, as in the genes for the haemoglobinopathies, to the totally culturally induced problems, such as the after-effects of clitoridectomy, the majority of conditions discussed are due to the multifactorial effects of genetic clustering and lifestyle differences. Discussion of osteomalacia, rickets and vitamin D deficiency provides an excellent example of the interaction and congruence of factors that result in 'ethnic' differences in disease incidence (Pacy 1989, Macbeth 1993). The reader's attention is also drawn to published collections of papers (e.g.

Cruickshank & Beevers 1989, McAvoy & Donaldson 1990) on ethnic differences in health conditions.

As the cultural control of conditions that lead to these biological differences is given greater recognition by clinicians and researchers, it is to answer their fundamentally biological questions about cause of illness that those from medical traditions seek cooperation from those well informed on ethnographies and cultural differences. Meanwhile those in the social sciences who can provide this kind of information are more concerned with answering quite different questions. The majority are not willing to apply their knowledge to the medical questions perhaps because they find these questions either uninteresting or daunting. Thus, while great hopes are sometimes pinned on biosocial dialogue, the result can be disappointing because of confused objectives. In the workshop from which this book emanated some participants expressed such disappointment. It was noticeable that many clinical geneticists simply wanted information to enrich their professional expertise, rather than to change it. They did not want to discuss the social significance of their *questions* or what the morality of their questions was. Meanwhile an impression given by some social scientists was that they considered themselves to be in some way the guardians of social morality, rather than dispassionate observers and recorders of it. Perhaps it is not inconceivable that the prime objective of some medical funds diverted to social scientists has been to predict and evade social criticism, or even to exonerate the medical professionals. However, it is unlikely that such an objective will succeed, for the resolution of ethical issues does not rest with any academic discipline; it lies in the wider public domain or in the individual. Biosocial dialogue is required on much humbler topics.

When Modell, in her contribution to the meeting and this volume, discusses the social organisation of kinship patterns, she is indicating an area where even a few weeks' education in social anthropology could clarify some information which is highly significant to the geneticist and so to the familial tracking of genetic risk. Similarly, geneticists could enrich their understanding through information on preferred and actual marriage patterns, how often this leads to inbreeding and how often consanguinity is suggested only on the basis of classificatory relationships. Other areas include the social processes that determine spatial and social mobility, the selectivity of migrants (Macbeth 1984) and, as discussed above, the attitudes to and consequences of physical and social barriers. On the other hand, gene frequencies can be used to dismiss the concept of a biological basis to the social claims of ancestral 'purity', and an understanding of multi-factorial causes of continuous biological diversity would be a useful factor in the social scientists' analysis of people's stereotyping of that diversity. I have suggested (Macbeth 1993) that cluster analysis of contemporary gene frequencies could even be used to 'test' the extent to which a claimed ethnicity had, in reality, affected past mating, as an examination of that ethnicity itself.

Simply because some of the difficulties and disappointments are highlighted, one should not lose sight of the gains from a biosocial workshop such as the one from which this book derives. Recognition of the difficulties is itself a gain. There have been notable advances in biosocial dialogue over a very few years and highly successful examples are when cross-disciplinary perspectives on a given topic are simply juxtaposed, leaving attempts at integration to the reader, as in the series of books sponsored by the Biosocial Society and published by Oxford University Press (e.g. Harrison 1988, Mascie-Taylor 1990, Macbeth 1996). Optimism about biosocial dialogue should be tempered, but not lost.

In conclusion, the biological anthropologists' study of biological diversity shows that ethnic groups are not only hard to define socially, but cannot be delineated genetically. Genetic comparison of groups can only be a comparison of frequencies of genes in people somehow classified into groups, but the discovery of any difference in such frequencies does not confirm the validity of the classification chosen. With the increasing mapping of genes which cause no more than a propensity for a multifactorial pathology, medical interest in genes and genetic screening is today not limited to the single gene traits. Yet multifactorial conditions are also affected by many other ethnic aspects of lifestyle which cannot be disentangled from each other or the genetic propensity. While self-identity of ethnicity affects marital patterns and so genes, it is logical to assume that self-definition of identity correlates even more absolutely with the social, economic and cultural practices that affect the multifactorial conditions than it does with the gene clustering. It is, therefore, important that 'proof' of genetic difference is not claimed on the basis of recording multifactorial cases. Understanding and vigilance are needed in order to ensure that comparative biological data do not become used as supporting evidence for the *social* concept of biologically discrete 'races'.

Self-definition of ethnic group remains our closest definition, even though self-definition can change not only with life circumstances but also with the context within which the definition is framed

(Macbeth & Bertranpetit 1995). With which group people (and before them their parents) identify will, in turn, affect, but not totally determine, their biologies, their marital choices and, in due course, the genes of their offspring.

NOTES

1. Gene frequencies are measured as the percentages of particular alleles of a given gene in a particular population.
2. As in Mendelian populations, discussed above.
3. Clinal distributions of gene frequencies arise where the proportions of alleles in adjacent subpopulations change gradually across the relevant geographical area.
4. Genoclines are clines (geographical gradients) in gene frequencies.

REFERENCES

Ardener, E. 1974 Social anthropology and population, in Parry, H. (ed.) *Population and its Problems*. Oxford University Press: Oxford, pp. 25–50.

Barbujani, G., & Sokal, R. 1990 Zones of sharp genetic change in Europe are also linguistic boundaries. *Proceedings of the National Academy of Sciences of the United States of America* 87: 1816–19.

Boyce, A.J., Kchemann, C.F., & Harrison, G.A. 1968 The reconstruction of historical movement patterns, in Acheson, E.D. (ed.) *Record Linkage in Medicine*. Livingstone: Edinburgh, pp. 303–19.

Chakraborty, R., & Weiss, K.M. 1982 Comments to N.E. Morton et al. *Current Anthropology* 23: 163–4.

Chapman, M.K. (ed.) 1993 *Social and Biological Aspects of Ethnicity*. Oxford University Press: Oxford.

Cruickshank, J.K., & Beevers, D.G. (eds) 1989 *Ethnic Factors in Health and Disease*. Wright: London.

Dobzhansky, T. 1962 *Mankind Evolving*. Yale University Press: New Haven.

Dolberg, O., Alkan, M., & Schlaeffer, F. 1991 Tuberculosis in Israel: a 10-year survey of an immigrant society. *Israeli Journal of Medical Science* 27(70): 286–9.

Garcia-Moro, C., Toja, D.I., & Bertranpetit, J. 1986 Análisis de la movilidad matrimonial en el Pallars Sobir (1918–1974). *Trabajos de Antropologia* XX (2): 221–37.

Gomila, J. 1976 *Définir la population*. In *L'Étude des isolats: Espoirs et limites*. INED: Paris, pp. 5–36.

Harris, H. 1980 *The Principles of Human Biochemical Genetics*, 3rd edition. Elsevier/North Holland: London.

Harrison, G.A. (ed.) 1988 *Famine*. Oxford University Press: Oxford.

66 Culture, Kinship and Genes

Kalter-Leibovici, O., Van Dyk, D., Leibovici, L., Loya, N., Erman, A., Kremer, I., et al. 1991 Risk factors of development of diabetic nephropathy and retinopathy in Jewish IDDM patients. *Diabetes* 40 (2): 204–29.

McAvoy, B.R., & Donaldson, L.J. (eds) 1990 *Health Care for Asians*. Oxford University Press: Oxford.

Macbeth, H.M. 1984 The study of biological selectivity in migrants, in Boyce, A.J. (ed.) *Migration and Mobility: Biosocial aspects of human movement*. Taylor and Francis: London, pp. 195–207.

Macbeth, H.M. 1993 Ethnicity and human biology, in Chapman, M. (ed.) *Social and Biological Aspects of Ethnicity*. Oxford University Press: Oxford, pp. 47–91.

Macbeth, H.M. (ed.) 1996 *Health Outcomes: Biological, Social and Economic Perspectives*. Oxford University Press: Oxford.

Macbeth, H.M., & Bertranpetit, J. 1995 Biology, boundaries and borders. *International Journal of Anthropology* 10: 52–61.

Mascie-Taylor, C.G.N. (ed.) 1990 *Biosocial Aspects of Social Class*. Oxford University Press: Oxford.

Morton, N.E., Kennett, R., Yee, S., & Lew R. 1982 Bioassay of kinship in populations of Middle Eastern origin and controls. *Current Anthropology* 23 (2): 157–67.

Mourant, A.E., Kopec, A.L., & Domaniewska-Sobczak, K. 1978 *The Genetics of Jews*. Oxford University Press: Oxford.

Pacy, P.J. 1989 Nutritional patterns and deficiencies, in Cruickshank, J.K., & Beevers, D.G. (eds) *Ethnic Factors in Health and Disease*. Wright: London.

Paine, R. 1989 Israel: Jewish identity and competition over 'tradition', in Tonkin, E., McDonald, M., & Chapman, M. (eds) *History and Ethnicity*. Routledge: London.

Rozen, P., Lynch, H.T., Figer, A., Rozen, S., Fireman, Z., Legum, C., et al. 1987 Familial colon cancer in Tel-Aviv area (Israel) and the influence of ethnic origin. *Cancer* 60 (9): 2355–9.

Stein, M., Kuten, A., Arbel, M., Ben-Schachar, M., Epelbaum, R., Wajsbort, R., et al. 1988 Carcinoma of the nasopharynx in Northern Israel: epidemiology and treatment results. *Journal of Surgical Oncology* 37 (2): 84–8.

Thomas, D.S. 1938 *Research Memorandum on Migration Differentials*. Social Science Research Council: New York.

Wright, S. 1943 Isolation by distance. *Genetics* 28: 114–38.

4 What is an Ethnic Group? The View from Social Anthropology

Ursula Sharma

In this paper I shall review some of the ways in which anthropologists have looked at and written about what we now call ethnic groups and ethnicity. I shall argue that the terms in which anthropologists have written about such issues can only to a very limited extent be regarded as a separate, let alone 'expert', form of discourse. I shall then examine the implications of this for any kind of dialogue between anthropologists and medical professionals.

IN THE BEGINNING...

In the beginning – or so anthropologists' contemporary accounts of the development of their discipline go – anthropologists assumed that the world, or at any rate that part of the world which was inhabited by 'primitive' peoples, was populated by more or less discrete groups. These were usually termed 'tribes', and whilst they did not necessarily exist in splendid isolation, they were nevertheless distinguishable from each other in terms of their culture and social structure. They might well differ in physique and appearance, but anthropological discourse by the 1930s had distinguished itself as a *social* anthropology and did not relate social and cultural difference to genetic difference. Individuals could be identified as belonging to one or other of these groups and, as members, acquired values through socialisation which led to profoundly held dispositions and attitudes (moral, aesthetic, political). To quote from one classic text:

> That every Nuer considers himself as good as his neighbours is evident in their every movement.... Nuer are most tenacious of their rights and possessions.... A child soon learns that to maintain his equality with his peers he must stand up for himself against any

encroachment on his person or property. (Evans-Pritchard 1940: 182–4)

Some of these groups were fairly isolated and even those who were not often had a strong sense of their own difference from their neighbours; to that extent the innocence of the early anthropologists is understandable. A less charitable critique has pointed out that the notion of 'tribe' is a particular kind of ideological understanding of the complexity and heterogeneity of populations brought together by the colonial state. It represented the way in which Europeans thought about these peoples, not necessarily the ways in which the colonised peoples thought about either themselves or each other.

Certainly the notion of the 'tribe' also belonged to a particular kind of administrative discourse. The policies of 'indirect rule' in Africa and the idea of 'not interfering with existing religious practices' in South Asia required a knowledge of prior social practices. In Africa it was expected that anthropologists could supply this knowledge. In South Asia administrators often went about collecting it themselves, resulting in the massive 'Tribes and Castes' volumes which were published in a number of provinces of British India in the late 19th century.

There has been much debate about the relationship between anthropology and colonialism (Asad 1973), which I will not rehearse now except to note that some commentators have suggested that anthropologists' uncritical acceptance of the division of populations into 'tribes', assumed to have some kind of primordial cultural unity, reinforced a colonial administrative discourse. Tribes and the consequent blight of 'tribalism' might, in fact, be as much a product of this discourse as its subject matter (see Godelier 1977: 96). And whilst anthropologists' notions of 'tribal' characteristics were based on socialisation rather than inherited traits, they did not, in practice, have the effect of undermining popular and administrative perceptions of tribal culture being associated with racial inferiority.

BOUNDARIES, ETHNICITY AND THE URBAN SCENE

This innocence, if innocence it was, did not last long. If ever people had lived in mono-ethnic isolation they patently did not do so in the post-war colonial world. Studies of urban Africans carried out in the late 1950s and '60s made it clear that, whilst most Africans did regard themselves as belonging to some kind of named ethnic group and were so identified by others, they also participated in many situations where 'tribal' values could not guide their behaviour and were even perceived as irrelevant. Putting the same development in another way, urban migrants developed a range of identities: worker (in a particular industry), African (as opposed to white), and membership in a tribe or ethnic group distinct from other tribes or ethnic groups (Epstein 1958).

Living in multi-ethnic locations or pluralist polities, however, does not in itself diminish the importance of ethnicity and cultural difference. The fact that people identify themselves as miners or textile workers does not prevent them from identifying themselves as Pakistanis or Sikhs. Indeed, the workplace may just as easily become a place where ethnic consciousness is heightened as one where occupational solidarities are formed. The key theoretical discovery of this period of anthropological endeavour was that ethnic groups are not 'naturally occurring' groups (as 'tribes' had been assumed to be) but the product of boundary construction and maintenance (Barth 1969; see also Jenkins 1986 for a more detailed account of this 'paradigm shift').

But boundaries are constructed by both·'them' and 'us'. Thus Okeley's study of the Gipsy Travellers renders irrelevant the notion that there is any use in trying to distinguish 'real' Gipsies from 'inauthentic' claimants to the title, though this is an idea to which both Gipsies and non-Gipsies often refer. The distinction between Gipsies and Gorgios (as Gipsies term those outside their group) is continually constructed by Gipsies and Gorgios alike. Group culture is treated as a resource for the construction of these boundaries, rather than a simple indicator independently determining where the boundaries are to be drawn. Thus Okeley treats Gipsies' appeal to notions of purity and pollution largely as a *means* of 'marking an important boundary with the outside social world', though she is clear that their sense of disgust at practices regarded as polluting is deep-seated and acquired in early childhood (Okeley 1983: 77ff).

Where the study of migrant groups in Britain was concerned, this theoretical development quickly took root. If some early studies of New Commonwealth immigrants were informed by the idea that cultural differences between these groups and the local indigenous population were the product of the fact that they had simply 'imported' unchanging cultural practices and attitudes from Pakistan or West Africa or wherever they had come from, anthropologists soon realised that cultural practices could not be studied solely in these terms (Watson 1977). Migrants themselves might well refer to 'traditions' which originated in a homeland located elsewhere, but 'imported' practices only constituted one component of their situation. This situation (which would include whatever economic niche they had managed to achieve, perceptions or prejudices expressed by the host population, policies of the local and national state) might encourage them to stress some practices, abandon others, develop new ones or give new meaning to old ones. Ethnic identity and culture was nothing if not dynamic.

This became particularly pertinent in the study of the 'second generation', who were often found to have quite complex notions of identity, often referring to different 'identities' in different situations. In a local oral history study in which I participated we found that people whose parents had come to Stoke-on-Trent from Pakistan, India or continental European countries did identify themselves as coming from 'the Potteries' in contradistinction from members of their ethnic group brought up in, say, Birmingham or Bradford or London, without denying a sense of being, for example, Pakistani or Polish (Sharma et al. 1989). This did not mean that their perception of what it meant to be a 'Potteries person' was the same as that of people whose parents had always lived in Stoke-on-Trent. Nor is 'ethnic identity' a straightforward or static concept. For instance, some girls of Pakistani parentage in Bradford were found to identify themselves increasingly in terms of being 'British Muslims' rather than 'Pakistani-British (Knott & Khokher 1993). Or ethnicity may be treated as a 'resource' which an individual may draw upon in some circumstances but not in others, depending on the extent to which appeals to ethnic solidarity, or to those links to which one's ethnic origins give one access, are regarded as more useful than other links and solidarities available to the individual. (Sandra Wallman's work on households in two areas of London exemplifies this approach: see Wallman 1984: 38ff.)

WHATEVER HAPPENED TO ETHNIC GROUPS? NEW VIEWS OF CULTURE?

The notion of identity as employed by anthropologists is not, of course, a new concept in social science (see Macdonald 1993: 7), having well-known antecedents, in psychology for instance. But does not its use in anthropology suggest an increasing tendency to see ethnicity not in terms of cultural *groups*, but in terms of a dimension of an *individual* sense of self? Are ethnic groups just categories of people who happen to share an ethnic 'identity', who may or may not cooperate as collectivities in maintaining this identity? And does ethnicity only exist where people of different 'ethnicities' interact? Is there no mileage left in the notion of ethnic boundaries as enclosing some kind of deep-seated cultural commonalities – people liking the same food, marrying according to the same rites, believing in the same values, and frequently putting pressure on each other to conform to shared norms?

Certainly some anthropological work concentrates so much on boundaries that it is difficult to tell whether there is any 'infilling'. (Do the boundaries enclose a group, and if so what kind of group and how is it held together?) However, anthropologists (and other social scientists studying minorities in this country) have been far from silent on this matter, though they have started to use new vocabularies. Whilst clearly rejecting the cultural essentialism of the earlier literature I have referred to, the idea that the behaviour of a Pakistani, a Sikh or a Jew is in some way to be simply 'read off' from the fact that he or she *is* a Pakistani, a Sikh or a Jew, anthropologists have not abandoned the collective dimension of their project entirely. Rather than conceiving of 'ethnic' groups as entities empirically defined by a common 'culture', anthropologists, in common with other academics who make culture the object of their study, are more likely to write in terms of cultural communities, communities which are not unified wholes, but are rather 'the locus of debate and divisions' (Banks 1996: 42ff, Weeks 1990: 94). We are dealing with groups of people who may or may not call themselves by the same name when called upon to fill in an ethnic monitoring form, but who share some kind of moral framework within which they can enter into dialogue upon issues which concern them. (What are the duties of a Muslim living in the British state? Is the notion of a beauty contest alien to Asian cultures? How can arranged marriages be conducted in a British context?) The essentialist notion of culture breaks down and the subcultural group which defines itself or is defined by others in terms of ethnicity turns out to be a chorus of voices.

MEANWHILE – OUT IN THE 'COMMUNITY'

Yet when it comes to public politics, where people belonging to different cultural communities must interact and enter into explicit or implicit contest with each other, there is widespread use (on the part of both majority and minorities) of a model which is much more like the older anthropological model and the tribal/ethnic group model used by colonial administrators. Ethnic groups are treated as groups which, even if no longer constituted by racial phenotypical difference, are empirically definable populations, sharing cultural values which are profoundly embedded in their consciousness, and sharing practices to which they are deeply attached (and which presumably are not easily changed, though open to erosion by threatening Others).

This set of assumptions is expressed in some forms of white racist discourse (the British are a race apart, they share certain values and characteristics which are endangered by the presence of large numbers of people of 'extraneous' origin who will never be able to learn or share them). Many ethnic minority groups will also, in certain situations, draw on a discourse which assumes that their characteristic cultural patterns are inherent, part of their 'heritage', something they are endowed with *qua* members of their ethnic group. Thus, some Hindus protested against the introduction of the poll tax on the grounds that it discriminated against joint family households with large numbers of young adults, a family form seen as part of Hindu 'traditional' culture.

In many public contexts, therefore, the terminology of ethnic groups predominates despite the 'deconstruction' I referred to in the last section. In spite of the criticisms of anti-racists, the discourse of multi-culturalism persists in the sphere of education, with its assumption that the school population is composed of children from various definable and monolithic 'cultural groups'. The race relations legislation enshrines a notion of a bounded 'ethnic group' which minorities have to go along with to some extent. Indeed, it has often been noted that ethnic minorities are obliged to draw on such discourses themselves, whether or not they reflect the realities of their situation. Rather like the colonial state, the modern state, at both national and local levels, has represented itself as managing the claims of different 'ethnic groups', balancing their demands on resources. More recently, with the demise of such local sources of funding as Urban Aid, it represents itself as managing and adjudicating moral or legal demands made in the name of ethnic groups of one kind or another (changes in the laws of blasphemy by British Muslims, claims of Sikhs to the right to wear turbans in school, etc.). Just as 'tribe' was an indispensable notion to British colonial administrators in Africa, so the notion of 'ethnic group' is indispensable to the modern British state both as an ideological means of understanding its own task and as an administrative means of controlling populations.

In this climate, culture becomes recognised as a means of resistance on the part of minority groups, a means of asserting identity and discovering collective strengths. But in the very act of doing so, the minority groups are also confronted with their own diversity. Multiculturalism had been opposed by an appeal to the notion of a single black people, constituted as a political group by both racist practices in the post-colonial metropolis and the class structure of capitalist society (Sivanandan 1982), regardless of their diverse origins. Such ideas were, for some time, very hard to question in public among those who had any pretensions to radical opposition to racism. Many academics also adopted the notion of a 'black' political category in their own studies (e.g. Runnymede Trust and Radical Statistics Group 1980; see especially the introductory note by Usha Prashar and Dave Drew).

Yet these ideas have recently been challenged by both activists and academics. For instance, Madood (an academic and activist) has stressed the value of group pride and identity and pointed out that this must often mean identification with a group less comprehensive than simply 'Black people'. There need be nothing divisive about encouraging different groups to oppose racism in the specific forms in which they encounter it and in terms of their specific cultural resources. 'For everybody is a somebody, not just a victim' (Madood 1990: 95). Ballard, an anthropologist, similarly recommends a greater acknowledgement of the cultural dimension of ethnicity on the part of those who are concerned for racial justice. Like Madood, he points out that the cultural traditions of different minority groups have often equipped them with specific resources contributing to their resistance to oppression, indeed their very survival (Ballard 1992). Elsewhere he has shown how the specific cultural resources which different migrant groups have brought to this country help to explain their different trajectories (Ballard 1990). The notion of cultural diversity has been rehabilitated, albeit in a very different sense from the old 'multiculturalism'.

Among academics, this acknowledgement of the diversity of the 'black experience' has not been confined to anthropologists. I could also refer to the contribution of sociology and cultural studies, though I have not done so since my brief was to write about anthropologists. To make amends, let me simply end this section of the paper with a quotation from work by Stuart Hall, a sociologist of Caribbean origin.

In the context of an appreciation of the diversity of Caribbean experience, he notes that:

> as well as the many points of similarity there are points of difference which constitute 'what we really are', or rather, since history has intervened, 'what we have become'. We cannot speak for very long with any exactness about 'one experience, one identity' without acknowledging its other side. Cultural identity is a matter of becoming as well as of being. It belongs to the future as well as to the past. It is not something which already exists, transcending time, history and culture. Cultural identities come from somewhere, have histories. But like everything which is historical they undergo constant transformation. (Hall 1990: 225)

DISCOURSES ABOUT 'ETHNIC GROUPS – ADMINISTRATIVE, ACADEMIC AND ACTIVIST'

In the course of this discussion I have identified three broad types of discourse about ethnicity and ethnic groups. First, there is what I have called the administrative notion of ethnicity employed by the colonial and post-colonial state. In this kind of discourse, which embraces the discourse of civil servants, politicians and many members of the professions, ethnic groups are treated as 'given' entities, relatively stable groupings of the population. Membership of such a group predicts the needs, rights and reactions of individuals. Second, we can identify the discourse of anthropologists and other social scientists attempting to analyse the nature of ethnicity and ethnic groups as a matter of professional academic interest and intellectual enquiry into the nature of social life. Whilst anthropologists come to this analytic task from a particular direction and with a particular type of theoretical baggage, the contribution of anthropology cannot be regarded as totally separate from that of sociology, cultural studies and various other disciplines. Third, we have the political discourse of activists and intellectuals among the ethnic minorities themselves, attempting to formulate their situation in such a way as to resolve the problems they see themselves as facing.

These categories, however, are extremely rough and far from discrete. Some social scientists are also activists, and some black intellectuals and professionals work for the local state. What has been called the 'race relations industry' stands at the intersection of all three. Indeed, the point of this paper has been to show how these discourses weave in and out of each other. Anthropological knowledge about ethnic groups is not something totally separate from the understandings which politicians and administrators hold, nor from the self-understanding of ethnic groups themselves. There is a limit to the extent to which anthropologists can be regarded as having some kind of 'expert knowledge' which is denied the rest of society, though no doubt some anthropologists would like to think that they do have such knowledge, and some certainly have very detailed knowledge about specific situations in which ethnicity manifests itself.

At the present moment the discourse of anthropologists about ethnicity and 'ethnic' cultures in this country has something in common with that of some black activists (their focus on culture as something which is constructed rather than 'given', their emphasis on identity, etc.), but there is nothing inevitable about this – in the past their discourse shared much more with that of administrators. Bourne and Sivanandan (1980) have accused anthropologists and sociologists of perpetuating these essentialist understandings of 'ethnic' culture. How far anthropologists are actually responsible for such understandings is an interesting question. Today's world is a reflexive one, and earlier definitions of culture by anthropologists have sometimes been taken on board by minority groups in their culture-building efforts (see Boddy 1993). Some anthropologists have certainly asked themselves whether focusing on ethnicity itself, however good the conceptual tools, may not fuel the spiral into divisive and antagonistic ethnicities and nationalisms which we see, in Europe today and, above all, in the Balkans. Is it not better that our (deconstructive) role be to 'expose the seductive simplicities which involve primordial loyalties and ethnic origins', rather than than to collude with the perpetuation of such simplicities (statement issued by Anthropologists Against Ethnic Violence, quoted in Benthall & Knight 1993: 20)

ETHNIC GROUPS, CULTURE AND MEDICINE

How might anthropological knowledge about ethnicity be relevant to health care professionals – in particular, to geneticists? This question cannot easily be answered until the professionals ask themselves why they might need to know about ethnic groups, and what they want to do with the notion. If they do this, and if the question is answered honestly, many different agendas must emerge. Some professionals are engaged in deflecting accusations of racism (medical schools and employing authorities). Nurses and midwives may be concerned to deliver a 'culturally sensitive' service to patients, or to understand behaviour in their patients which they find puzzling. Health educators and GPs may be concerned to secure better compliance with accepted medical wisdom about nutrition, preventive care, safe sex, etc. Broadly, most of these concerns locate medical discourse about ethnicity in the category I have called 'administrative discourse' and, in common with other instances of this genus, call for shorthand descriptions of ethnic groups which identify members and 'explain' or 'predict' behaviour. Accounts of ethnicity and ethnic grouping not delivered in this form are not easily assimilated into this kind of discourse.

Those called upon to deliver this kind of stereotypical information, however – anthropologists, members of ethnic minorities serving to represent their groups, and others – are increasingly unwilling to do so. Partly, this unwillingness is related to the abandonment of a determinist view of ethnic culture, as already described earlier in this paper. The misgivings also relate to the use to which such information is put. There has been much concern that a focus on the culture of ethnic groups, rather than on the racism and other disadvantages they experience in this country, simply leads to victim blaming. Uncritical use of a determinist view of ethnic culture has led to, for instance, the blaming of Asian cultural practice – or more correctly, Asian practice as perceived or conjectured by white professionals – for the high incidence of rickets among Asian women in Glasgow. Once health professionals have accepted a determinist view of ethnic culture as automatically giving rise to particular behaviours, this tendency is hard to reverse. It has certainly led to a pathologising of ethnic cultures rather than attempts to learn from them, and to a neglect of the material factors stemming from class or racial disadvantage which might account for differences in morbidity between ethnic groups (see Ahmad 1992).

For health care professionals, it seems that cultural practice, genetic inheritance and self-identity do not appear to be so readily separable in the everyday clinical context as social science discourse now proposes. First-cousin marriage appears to be far more common among Muslims of Pakistani ancestry than among other ethnic groups. Whilst not all 'Asians' practise arranged marriages, it is uncommon among people of white ethnic English parentage. In Britain, Tay-Sachs disease is uncommon outside certain Ashkenazi Jewish communities, and thalassaemia is unusual among people who do not come from, nor have a parent or ancestor from, the Eastern Mediterranean, Pakistan or North India. These are common-sense observations, and those who attempt to deliver appropriate health care services in an ethnically diverse community will not perceive it as helpful to be told that cultural practice, genetic inheritance and ethnic self-identification should be treated as three quite distinct dimensions of difference. The community nurse, the health visitor and the genetic counsellor engage with clients and patients on the very ground where these dimensions of difference appear to overlap and interact.

There are no easy answers to this problem. Rapp (1993) is correct to suggest that anthropologists may draw on their knowledge of the situation of minority groups to help health care professionals to 'pluralise' their services so that, for instance, amniocentesis and genetic counselling are provided as a 'service to [many] women' rather than to 'a [standard] woman'. Yet I have already suggested that anthropologists should not be regarded as having some kind of objective 'expert knowledge', quite distinct from other forms of knowledge and derived from some kind of totally neutral standpoint independent of the rest of society, which can simply be appropriated and applied by health care professionals or policy-makers to solve this problem. (Anthropologists may, of course, in some cases be well placed to reiterate what ethnic activists themselves have said, when the latter have spoken out but not been listened to, but this is a different matter).

Probably the best use of anthropologists' knowledge here can be to attend to their analyses of how ethnicity interacts with other dimensions of difference. Rapp (1993) has shown that genetic counsellors need to appreciate how their clients may not share the same medical perspective as the genetic counsellor when considering the 'risks' that their genetic histories might seem to carry for their unborn children. But neither can their approach to such issues be understood solely in terms of ethnic cultures. She gives the example of Mari-Carmen, a Puerto Rican interviewee in New York, who went along with her Pentecostalist husband's insistence that their unborn child would be protected by the infant Jesus whatever health problems it might have inherited. The woman herself was far more worried about her other two sons, growing up in a very rough neighbourhood, and about her husband, who had been involved in drugs (Rapp 1993: 187).

We might add that, except in so far as access to good schooling, housing and employment may be different for different ethnic groups, her responses could not be understood simply in terms of her 'ethnic culture'.

Without denying the role of ethnic cultures, therefore, we can show how the concrete life circumstances of minority groups affect the ways in which they prioritise their health care needs, and consequently the way they use services and what they expect from them. This is particularly important because, in my experience, the perception of particular ethnic groups by health care professionals is often based on observations in a particular district, where the members of the ethnic group in question may indeed be locally concentrated in certain jobs and housing, their class position consequently being fairly homogeneous. Yet even within a restricted locality, the experiences of members of the same ethnic group may be strongly differentiated along class and gender lines, as Khanum has shown in the case of a highly concentrated Bangladeshi group (Khanum 1994). Health care professionals need to be conscious of the ways in which Bangladeshi women and men, workers, entrepreneurs and unemployed may bring different experiences and needs to the clinic, yet remain united by certain shared 'Bangladeshi' experiences and background.

The view from anthropology, therefore, is not that we must abandon the notion of 'ethnic groups', but that our knowledge of them only yields practical results when we are prepared to abandon the notion of ethnic groups as discrete, bounded and homogeneous cultures. Instead, we need to accept ethnicity, as treated in the various discourses I have discussed, as one of a variety of factors which together produce the plurality of identifiable patterns of dispositions, constraints and choices seen in the complex urban culture of contemporary Britain.

REFERENCES

Ahmad, W. (ed.) 1992 *The Politics of 'Race' and Health*. Race Relations Research Unit, Bradford and Ilkley Community College.
Asad, T. 1973 *Anthropology and the Colonial Encounter*. Ithaca Press: London.
Ballard, R. 1990 Migration and kinship: the differential effect of marriage rules on the process of Punjabi migration to Britain, in Clarke, C., & Vertovec, S. (eds) *South Asians Overseas*. Cambridge University Press: Cambridge.

Ballard, R. 1992 New clothes for the emperor? The conceptual nakedness of the race relations industry in Britain. *New Community* 18(3): 481–92.

Banks, M. 1996 *Ethnicity: Anthropological Constructions*. Routledge: London.

Barth, F. 1969 *Ethnic Groups and Boundaries*. Allen and Unwin: London.

Benthall, J. & Knight, J. 1993 Ethnic alleys and avenues. *Anthropology Today* 9(5): 1–2.

Boddy, J. 1993 Managing tradition: 'superstition' and the making of national identity among Sudanese women refugees. Paper delivered at the Association of Social Anthropologists Decennial Conference, Oxford.

Bourne, J., & Sivanandan, A. 1980 Cheerleaders and ombudsmen: the sociology of race relations. *Race and Class* XXI (4): 331–52.

Epstein, A. 1958 *Politics in an Urban African Community*. Manchester University Press: Manchester.

Evans-Pritchard, E. 1940 *The Nuer*. Oxford University Press: Oxford.

Godelier, M. 1977 *Perspectives in Marxist Anthropology*, trans. R. Brain. Cambridge University Press: Cambridge.

Hall, S. 1990 Cultural identity and diaspora, in Rutherford, J. (ed.) *Identity, Community, Culture, Difference*. Lawrence and Wishart: London.

Jenkins, R. 1986 Social anthropological models of inter-ethnic relations, in Rex, J., & Mason, D. (eds) *Theories of Race and Ethnic Relations*. Cambridge University Press: Cambridge.

Khanum, S. 1994 We just buy illness in exchange for hunger: experiences of healthcare, health and illness among Bangla Deshi women in Britain. Ph.D. thesis submitted to the University of Keele (Department of Sociology and Social Anthropology).

Knott, K., & Khokher, S. 1993 Religious and ethnic identity among young Muslim women in Bradford. *New Community* 19(4): 593–610.

Macdonald, S. (ed.) 1993 Introduction to *Inside European Identities*. Berg: Oxford.

Madood, T. 1990 Catching up with Jesse Jackson: being oppressed and being somebody. *New Community* 17(1): 85–96.

Okeley, J. 1983 *The Traveller-Gipsies*. Cambridge University Press: Cambridge.

Rapp, R. 1993 Amniocentesis in sociocultural perspective. *Journal of Genetic Counselling* 2(3): 183–96.

Runnymede Trust and Radical Statistics Group 1980 *Britain's Black Population*. Heinemann: London.

Sharma, U. (ed.) 1989 *Putting Down Roots: Some life stories from the Potteries*, collected by Talking Back Collective. Staffordshire Multi-cultural Advisory Service: Stafford.

Sivanandan, A. 1982 *A Different Hunger: Writings on black resistance*. Pluto Press: London.

Wallman, S. 1984 *Eight London Households*. Tavistock: London.

Watson, J. 1977 *Between Two Cultures: Migrant minorities in Britain*. Blackwell: Oxford.

Weeks, J. 1990 The value of difference, in Rutherford, J. (ed.) *Identity, Community, Culture, Difference*. Lawrence and Wishart: London.

Part II
Genetic Disease in Context

5 Consanguineous Marriage and Genetics: A Positive Relationship

Aamra Darr

This volume, and the conference from which it has arisen, both demonstrate and reflect the need for cross-disciplinary research and practice in health care generally, and genetics in particular. The traditionally imposed perspectives of social scientists and health professionals do not encourage the mutual recognition of the other's relevance. Two necessary criteria for successful cross-disciplinary research initiatives are i) the involvement from the very beginning of all relevant disciplines in the development of any collaboration, so that the research agenda takes into account all necessary perspectives, and ii) adequate funding to permit the exploration of emerging concepts and situations.

The purpose of this chapter is to relate some of the findings of a research project that fulfilled these criteria and which clearly demonstrates how insights derived from a social science perspective can contribute to the practical application of health care services. This project (Darr 1990) explored the experiences of British Pakistani families with children affected by thalassaemia, and assessed their implications for service delivery to this population. The research also analysed the social and medical implications of consanguineous marriage, a highly emotive issue riddled with misunderstandings among health personnel as well as communities practising this marriage pattern, including the British Pakistanis. These misunderstandings have serious consequences for the development and delivery of health care to populations in need of that care (Darr 1990, Bittles et al. 1991, Modell & Kuliev 1992). Ahmed's pertinent analysis and assessment of the 'consanguinity hypothesis' concludes that, 'the fascination with consanguinity in the NHS owes more to racism than to science' (Ahmed 1995: 68).

The main focus of this chapter will be an illustration of how the practice of consanguineous marriage, viewed as a problem by many health professionals, has positive relevance for the delivery of a genetics service to the British Pakistani population. This approach is derived from an anthropological analysis of the (British) Pakistani kinship system. The networks formed by the kinship pattern provide an opportunity to offer a genetic counselling service by utilising these very networks. A model for genetic health service delivery to the British Pakistanis is proposed which can be applied not only to thalassaemia but also to other genetic diseases.

The chapter is organised in three sections. The first provides the background to the research project, placing it in context and explaining the choice of this subject. The second section describes the components of the kinship structure of British Pakistanis, a knowledge of which is essential for understanding the development of an appropriate delivery model for genetics services. The final section sets out the proposed model of genetics services, and concludes with comment on the approaches employed for providing genetic counselling to British Pakistanis and, more generally, the dissemination of genetic information to all minority ethnic populations.

THE STUDY

The study of the experiences of Muslim British Pakistani families with children affected by thalassaemia was initiated by Bernadette Modell, who had noted a particularly low uptake rate of prenatal diagnosis in this population (Modell et al. 1985), and was conducted over the years 1981–86. Many paediatricians had also expressed concerns about the increasing numbers of children with thalassaemia, their difficulties in communication with the children's parents, and what they perceived as the parents' negative attitudes towards the management of the disease and towards prenatal diagnosis. Another stimulus to carry out the study was the stated view of some members of the medical profession that it was pointless discussing the option of prenatal diagnosis with Muslims because they would not accept terminations of pregnancy. A widespread belief among health professionals, that the consanguineous marriage pattern of this population was a major determining factor in its 'genetic health', also required further investigation.

It was clear to Modell that technological improvements alone were insufficient for dealing with thalassaemia comprehensively, and a sociological/anthropological approach to the study of the problem was necessary to define appropriate approaches for the effective delivery of the medical services. It would be necessary for the researcher to be a female social scientist of the same religious, cultural and linguistic background as the families and who identified with her own culture and people. Because the research project entailed discussion of personal issues such as childbirth, it was imperative that the researcher was of sufficient maturity to deal with these and related issues. Only when a researcher with the necessary attributes was found did the study commence. The separate agendas of the clinician and the social scientist could have led to misunderstandings and even conflict, and so continued discourse was an integral part of the project. The discourse enabled an ongoing assessment of the research methods employed and the implications of emerging issues as the project developed.

The central objective of the research was the identification of the most appropriate approach to carrier diagnosis and genetic counselling for thalassaemia major among British Pakistanis. The fieldwork, carried out between 1981 and 1986, began in a city (city Z) with a high concentration of British Pakistanis; as it progressed, the research involved other British Pakistani families in different parts of England. It began by exploring the experiences of patients and families. Contact was established with a total of 31 families, including 45 affected children. Eleven families were studied intensively and another five families were studied less comprehensively. The interactions between myself (a female Muslim British Pakistani) and the families took place in their homes in our mother tongue. This was a departure from the families' usual meetings with health professionals – involving male doctors in a hospital setting, the exchanges being conducted in English. From virtually the first contact with a family in their home, the meetings became counselling sessions. This was the first opportunity the parents had often had to speak about their child and the disease with someone who spoke their language, gave them ample time, and understood their beliefs, apprehensions and frustrations. The experiences were written up as case histories at the end of the research period, incorporating the families' changes in attitude over time resulting from developments in technology and from family experiences.

One of the most striking features during the initial visits was the isolation of the families resulting from their lack of information about the disease and their lack of contact with other affected families. In city Z, 18 children from 14 families were treated at two different hospitals and by different doctors. Only a few couples knew of other children with thalassaemia in the city, whilst most parents thought their child was the only one with the condition. This isolation of the families made it clear that a group meeting open to all the families had to be organised. Some couples had already expressed a desire to meet others in a similar situation. When consulted, every family wished to meet.

A support group was formed, organised and run by myself, where doctors were invited to speak and parents were able to exchange experiences. The group was very successful. Despite only very few families owning cars, parents organised their own transport; husbands and wives came regularly, as well as children, some accompanied by relatives or friends. Couples began offering support to one another, as individual families visited each other to discuss issues pertinent to their children's condition, such as prenatal diagnosis, bone marrow transplantation, the controversy surrounding consanguineous marriage and their experiences of treatment in Pakistan. The success of the group stimulated interest in nearby towns, with the result that either a support group was formed or a meeting of parents was organised. In city Z, soon after the families had come together, all the patients were brought under the care of one paediatrician and standardised treatment was provided. For the rest of the study, the group meetings became an integral part of the families' involvement with the disease and its management.

The willingness of British Pakistanis to participate in group activities related to the disease was a result both of their desperate need to unburden themselves of the anxiety caused by isolation, and also of their desire to gather more information about the condition and how to cope with it through sharing their experiences. The latter, particularly, is a reflection of the relative lack of stigma attached to the disease. This recognition by British Pakistani couples of the value of support groups is an important component of the proposed model for the delivery of a genetics service to this population.

Relevance of Kinship Patterns for Inherited Disorders

An interesting and unexpected feature of the research was the speed with which information passed between families in different cities. Despite the limited scale and duration of this project, families in other parts of England became aware of the developments in city Z and the adjoining areas through informal family networks. As the researcher, I was contacted directly by families from two different cities who were unconnected with the research. In addition, visits took place between distant kin, when families in city Z met relatives from different cities to discuss their experiences of bone marrow transplantation and prenatal diagnosis. Information was evidently being exchanged through pre-existing, established kinship networks.

This development led to a closer examination of the kinship pattern of British Pakistanis and of consanguineous marriage, which is an integral part of that pattern. The ensuing analysis led to the conclusion that *knowlege of the kinship patterns of different communities is essential in devising appropriate services for genetic disorders*, as varying kinship and marriage patterns reflect the existence or non-existence of networks which can be utilised for genetic counselling and for the mutual support of families with affected children.

COMPONENTS OF THE (BRITISH) PAKISTANI KINSHIP STRUCTURE

Consanguineous Marriage

Consanguineous marriage is a highly favoured practice among Pakistanis and British Pakistanis. In a study of married couples, 83 per cent of partners were related to each other, with 55 per cent being first cousins. In the preceding generation, the wife's parents were related in 70 per cent of couples, with 31 per cent being first cousins (Darr & Modell 1988). The latter figures are in accordance with findings for the same generation in Pakistan, where an enquiry among 900 couples showed 73 per cent to be related, with 36 per cent being first cousins (Shami & Zahida 1982). The figures reflect a shift in the younger generation – the parents of the affected children in this study – towards marrying relatives who are more closely related. This creates a multiplicity of relationships within and between large, extended families, as indicated in Figures 5.1 and 5.2.

The Household

The primary social unit in Pakistan and Britain is the household. The household may be of many types. Within the household each position:

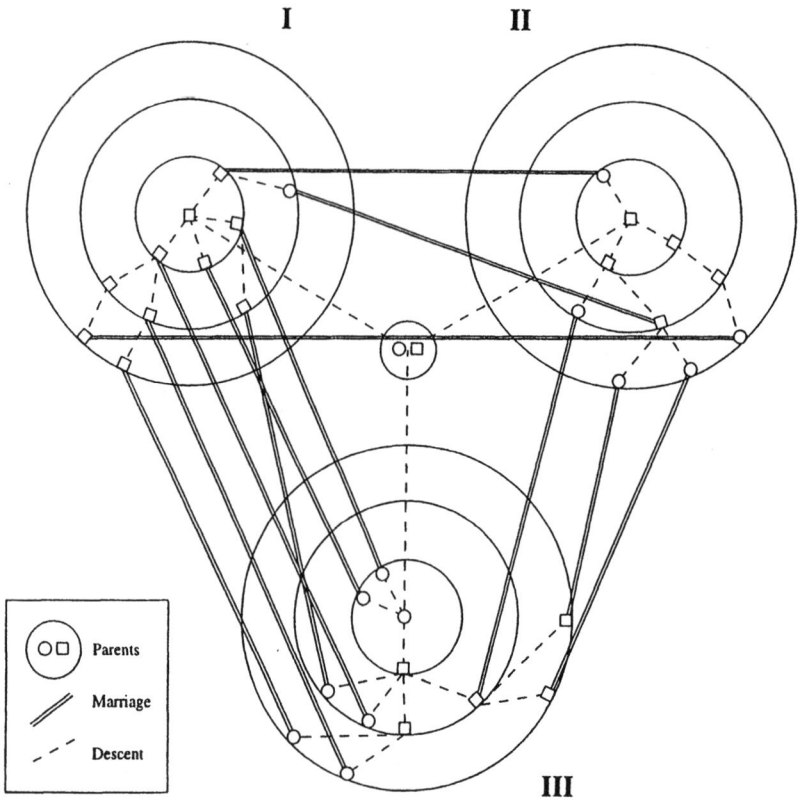

Figure 5.1   Intermarriage among three families.
Source: Eglar 1960: 208

Comprises a complex of rights and duties, attitudes, expectations
and sentiments which are balanced carefully to ensure the effective
functioning of the unit. Roles are precisely defined into an inter-
locking pattern of mutual interdependence and individual subordi-
nation to the group. (Saifullah-Khan 1977)

From his study of British Pakistanis in Rochdale, Anwar (1982)
cites examples of households in which several brothers, at least one of
whom will be married, live with their wives and children and sometimes with their parents.

Among British Pakistanis there are fewer extended families
than in Pakistan. The households tend to be of two generations
rather than the three or more found in Pakistan (Anwar 1985).

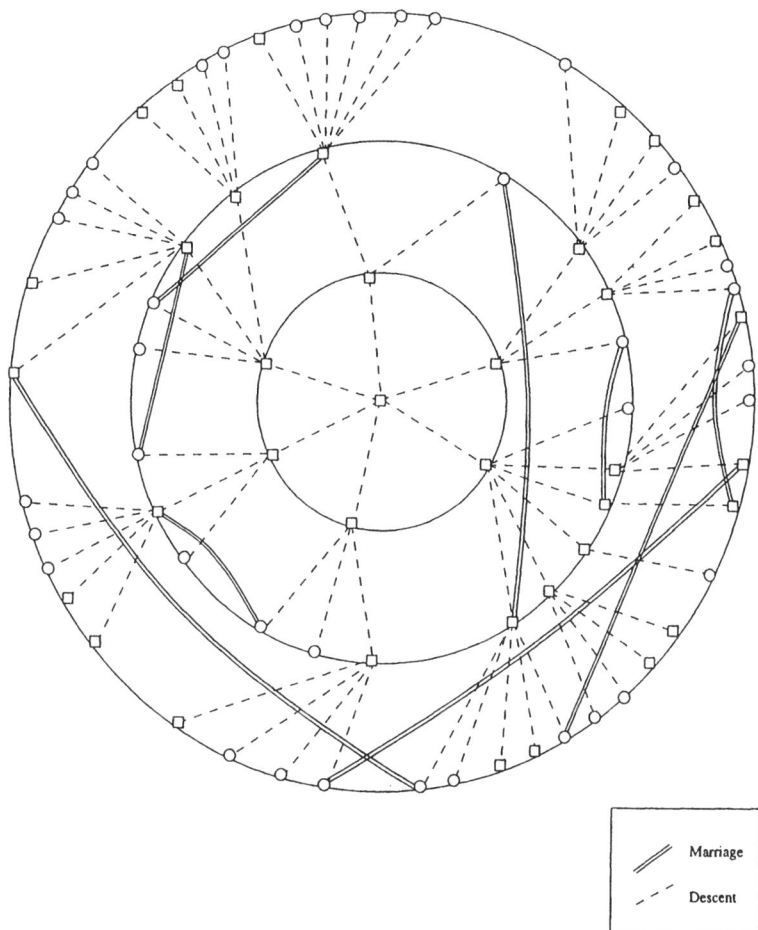

Figure 5.2 Intermarriage in one family.
Source: Eglar 1960: 209

A national study (Anwar 1978) found that of 549 Muslims interviewed (mostly of Pakistani origin), 71 per cent were living in nuclear families and 29 per cent in extended families. Nine of the 11 intensively studied families in the present study of families with thalassaemia lived as nuclear families. The other two went through stages of living as nuclear families and as extended families, as dictated by financial and other circumstances.

The *Biradheri*

It is the kinship system which regulates and structures relationships beyond the household, and here the basic institution is the *biradheri* (Alavi 1972). The *biradheri* (brotherhood) is an endogamous group whose members claim descent in the paternal line from a common male ancestor. Related to the effective functioning of the *biradheri* is the practice of *vartan bhanji* which is a mechanism of gift exchange. The term *vartan bhanji* literally means 'dealing in sweets', but in social terms it means 'dealing in relationships'. The transactions which give this practice its name take place on occasions such as a birth, the circumcision of a son, marriage and death, as well as other significant occasions. Apart from the exchange of gifts, other examples of *vartan bhanji* would be offering someone an interest-free loan for the deposit on a house, or providing help with the cost of a marriage. The system gives financial support to *biradheri* families in times of considerable expenditure, and expresses official alignment and solidarity with other households. Alavi (1972) states that the exchanges are:

> transactions in a perpetual cycle of regular reversal of ritual debts between the households. Each payment, notionally, consists of two parts; one part extinguishes the pre-existing debt and another part creates a new debt in the reverse direction.

The *biradheri* functions as a welfare, financial and advice service. In Pakistan it extends beyond the village to other villages in the locality, and closely related *biradheri* members who live at long distances remain in close touch and attend important family functions. *Biradheri* networks also extend outside Pakistan, taking in members of their *biradheri* in Britain and in other countries where Pakistanis have settled. Anwar (1985) notes that:

> It is often difficult to draw the line at where the Biradheri ends, as this depends on the person defining those he thinks belong to his Biradheri. In practice it depends on the contact maintained with each other, the degree to which consanguinity is practised, the frequency with which members meet to take decisions which affect the Biradheri as a whole, or participate in ritual ceremonies and in Vartan Bhanji.... Distantly related consanguines, those related affinaly, or friends who have assumed either affiliation due to regular rituals or certain kinds of exchange of services may be

incorporated into one's Biradheri. This type is usually called 'fraternal Biradheri'.

The *Quom*

A wider and much looser network of affiliation than the *biradheri* is the *quom*, which still has some relevance in terms of choosing marriage partners:

> Although the caste system is rejected in Islam there are clear vestiges of the pre-partition social structure. Notions of purity and pollution, restricted commensality and certain other features of the caste system are less evident in Mirpur and Pakistan but there is a general hierarchy of castes (with land-owning castes at the top and service castes lower on the scale). Not all sons follow their father's traditional occupation (e.g. 'Lohar'-blacksmith, 'Nai'-barber, etc.) but the caste ('quom') name is retained and is thus significant in certain situations. (Saifullah-Khan 1977)

The institutions of *biradheri* and *vartan bhanji* are firmly entrenched in the social fabric of Pakistani society, whether in rural or urban Pakistan or in Britain. In each setting, affiliations are made and solidarity among kin is maintained in virtually identical ways; networks are formed within which the processes of obligation, support and control can operate. Figure 5.3 shows the major institutions constituting the Pakistani kinship system which have been extended to and established in Britain, indicating the local and international span of the kinship networks.

These networks were the channels in use when information generated in city Z began to be transmitted swiftly between cities in England.

A POSSIBLE APPROACH TO GENETIC COUNSELLING FOR THE BRITISH PAKISTANI POPULATION

Due to the consanguinity, recessively inherited conditions tend to be clustered in family groups rather than scattered at random throughout the population, as would be found in a population with a non-consanguineous marriage pattern.

Cities in Pakistan

Villages in
Pakistan

Cities in Britain

Household
Biradheri
Quom

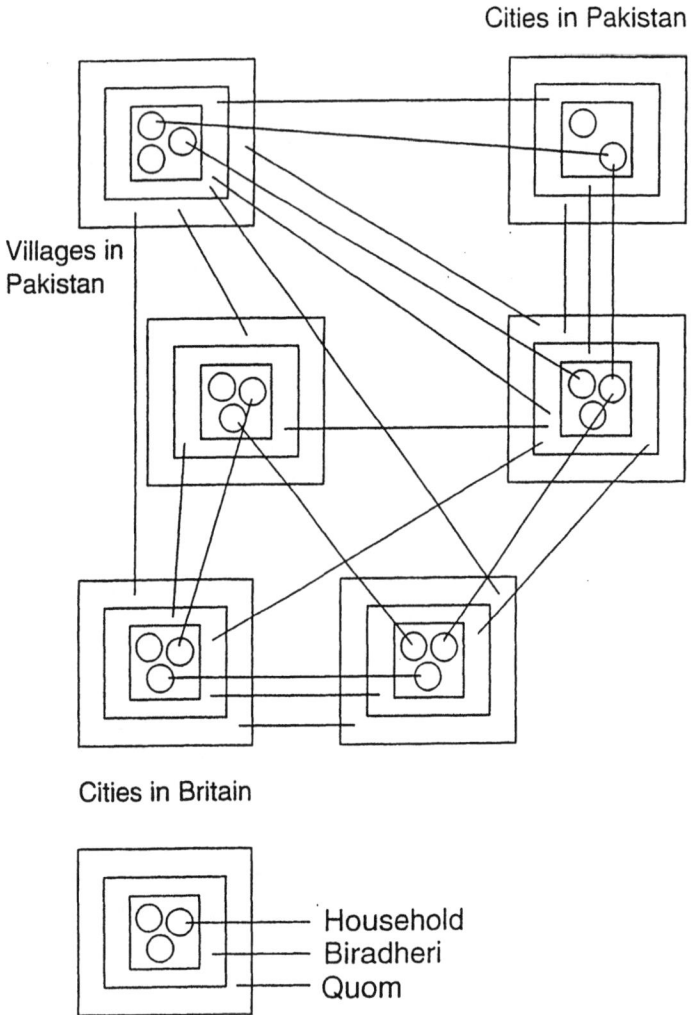

Figure 5.3 Kinship networks of (British) Pakistanis.

The birth of an infant with a recessively inherited disorder identifies a family that is transmitting the gene. When consanguineous marriage is common and family size is large, as in the British Pakistani population, most recessively inherited conditions will come to attention with

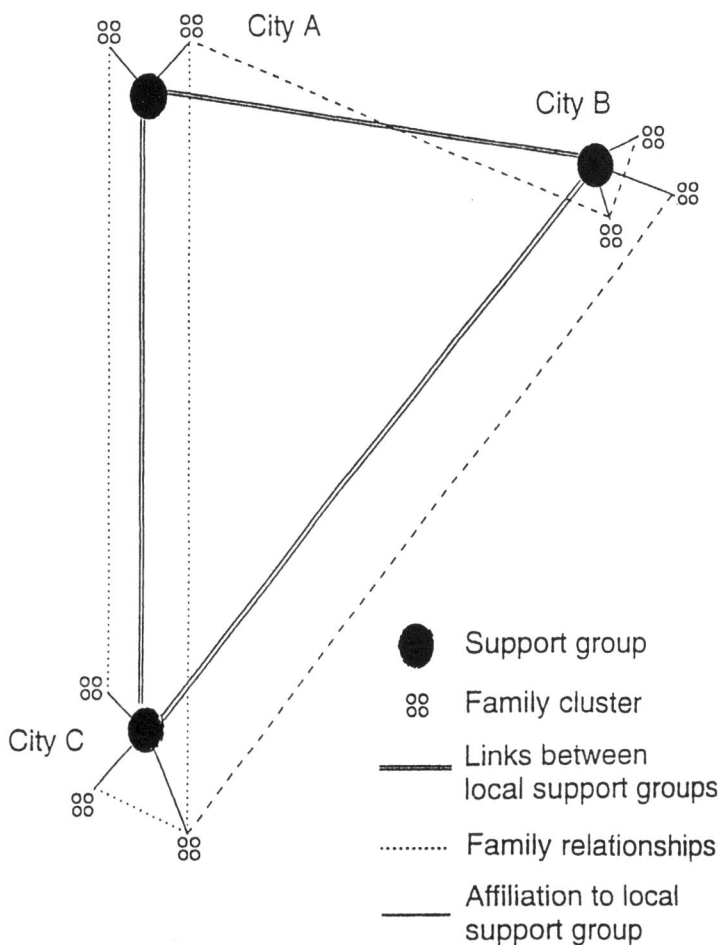

Figure 5.4 Model showing possible networks of information and support formed through genetic counselling and support groups.

the birth of an affected child somewhere within a family cluster. Genetic counselling focused on the extended family of affected infants may then be particularly successful in detecting couples at risk of having affected children and advising them prospectively.

A model of possible networks of information and support formed through genetic counselling and support groups is given in Figure 5.4.

This model is based on the use of existing social networks which are peculiar to a population that has a relatively high frequency of consanguineous marriage. This provides an opportunity to offer a genetic counselling service in a unique way by making positive use of these social networks.

Using each family with a homozygote as a starting point, the genetic counsellor can counsel the family of the homozygote and then – with appropriate sensitivity – extend the counselling in a cascade fashion to the siblings of the parents and further out into the family. In this way, genetic counselling can be provided for a cluster of families with a known high genetic risk. Alongside the work with family clusters, the establishment of support groups could bring together the various families with already affected children, creating further networks between families where information and support could flow.

As rare conditions are even more likely to be limited to family clusters, the same approach could be applied to families with conditions for which carrier testing is not available. In this study, two of the families were also transmitting the gene for cystic fibrosis and one family was transmitting the gene for 'maple syrup urine' disease.

For the detection of those carriers of thalassaemia who are scattered around the country rather than being concentrated in particular cities and areas, it is imperative that the genetics services for the wider population (routine screening at antenatal clinics, GP surgeries, Well Woman clinics, etc.) are responsive to the needs of a culturally heterogeneous population.

In Britain, populations with different marriage conventions live side-by-side and intermingled, and consequently there are differing patterns of manifestation of inherited conditions. There is, however, a single health service responsible for the delivery of a genetics service to the whole population. In North-West Europe, including the UK, there has been only limited success so far in providing a genetics service for disorders that affect diverse and unevenly distributed ethnic minorities (Royal College of Physicians 1989). The specific pattern of manifestation of recessively inherited diseases among British Pakistanis can, however, be used to advantage in developing and delivering a highly effective and sensitive genetics service to this population, with only minor adaptations to the present service.

This approach is not only relevant to thalassaemia among British Pakistanis, but can be applied to all genetic diseases in populations which favour consanguineous marriages, and is being incorporated into World Health Organisation advice on policy development for the delivery of genetic health services in the Middle East. In Pakistan, steps have been taken by a team of health professionals at the Armed Forces Institute of Pathology, Rawalpindi, to implement this model of genetics services in collaboration with the Fatimid Foundation and the Pakistan Thalassaemia and Welfare Society (personal communication). The overall findings of the research have also been used in the ethnic minority section of a report of the Royal College of Physicians (1989).

CONCLUSION

The proposed approach to imparting genetic information and counselling in families is non-intrusive, as it begins by dealing with those families in the greatest need of genetics services, and they are seen in their homes and in their own cultural context. By focusing on the area of need and diffusing outwards, rather than on the community as a whole and working in towards specific families, it is non-alarmist. The proposed strategy avoids the time, expenditure and potential pitfalls of a health education campaign, as favoured by some health professionals, which would try to impart complex genetic information to large numbers of people; a scanty knowledge of genetics may do more harm than good. In Britain, where racial tensions already exist, it is imperative that health care initiatives do not single out ethnic minority groups through advertising the health problems to which particular groups are susceptible. This is especially important when relevant social customs, such as consanguineous marriage, may provoke prejudiced reactions. Information on inherited conditions must be presented in the wider context of different genetic conditions affecting every population worldwide. Until it is well established and widely known that all populations have to contend with inherited diseases, any initiative in genetic screening and counselling should be treated with extra sensitivity and care.

REFERENCES

Ahmad, W. 1995 Consanguinity and related demons: science and racism in the debate on consanguinity and birth outcome, in Samson, C., & South, N. (eds) *The Social Construction of Social Policy*. Macmillan: London.

Alavi, H. 1972 Kinship in West Punjab villages. *Contributions to Indian Sociology* 6: 1–27.

Anwar, M. 1978 *Between Two Cultures*, 2nd edition. Commission for Racial Equality: London.

Anwar, M. 1982 *Young People and the Labour Market.* Commission for Racial Equality: London.

Anwar, M. 1985 *Pakistanis in Britain: A sociological study.* New Century Publishers: London.

Bittles, A.H., Mason, W.M., Greene, J., & Rao, A.R. 1991 Reproductive behaviour and health in consanguineous marriages. *Science* 252: 789–94.

Darr, A.R. 1990 The social aspects of thalassaemia major among Muslims of Pakistani origin in England: family experience and service delivery. Ph.D. thesis, University College, London.

Darr, A.R., & Modell, B. 1988 The frequency of consanguineous marriage among British Pakistanis. *Journal of Medical Genetics* 25: 186–90.

Eglar, Z. 1960 *A Punjabi Village in Pakistan.* Columbia University Press: New York and London.

Modell, B., & Kuliev, A.M. 1992 Social and genetic implications of customary consanguineous marriage among British Pakistanis. *Galton Institute Occasional Papers*, second series, no 4. The Galton Institute: London.

Modell, B., Petrou, M., Ward, R.H.T., Fairweather, D.V.I., Rodeck, C., Varnavides, L.A., & White, J.M. 1985 Effect of fetal diagnostic testing on the birth rate of thalassaemia in Britain. *Lancet* II: 1383–6.

Royal College of Physicians 1989 *Prenatal Diagnosis and Genetic Screening: Community and service implications.* Royal College of Physicians: London.

Saifullah-Khan, V. 1977 The Pakistanis: Mirpuri villagers at home and in Bradford, in Watson, J.L. (ed.) *Between Two Cultures: Migrants and minorities in Britain.* Blackwell: Oxford.

Shami, S.A., & Zahida 1982 Study of consanguineous marriages in the population of Lahore, Punjab, Pakistan. *Biologia* 8: 1–15.

6 Factors Associated with Birth Outcome in Bradford Pakistanis

Sue Proctor and Iain Smith

INTRODUCTION

It is widely recognised that babies born in England and Wales to women of Pakistani origin have higher stillbirth (SBR) and perinatal mortality rates (PNMR) when compared with those from other ethnic groups (Terry et al. 1980, Gillies et al. 1984, Chitty & Winter 1989, Bundey et al. 1989). In 1991 the SBR for babies whose mothers were born in Pakistan was 7.9 per thousand (total births) compared with 4.4 per thousand (total births) to all women born in the United Kingdom (Office of Population Census and Survey 1993). Babies with Pakistani mothers also have a high incidence of lethal congenital abnormalities. In Birmingham it has been estimated that 41 per cent of infant deaths in Pakistani families were due to congenital abnormalities (Bundey & Alam 1993).

The reasons for the relatively high rates of poor birth outcome in this group remain unclear. It is clear, however, that a number of known confounding variables operate and can be summarised as shown in Figure 6.1.

Consanguinity

Previous studies of ethnicity and birth outcome both in Bradford and elsewhere have paid particular attention to consanguinity as the principal contributory factor (Lumb et al. 1981, Ahmad et al. 1989, Darr & Modell 1988, Baker et al. 1988). It has been estimated that the rate of consanguineous marriage among Bradford Pakistanis is approximately 60 per cent, which is similar to that reported from Birmingham and higher than in parts of Pakistan (Darr & Modell 1988). It has been reported that Pakistanis have an increased birth prevalence of autosomal recessive conditions (Young 1990) and of some serious conditions which have a partial genetic basis, such as neural tube defects

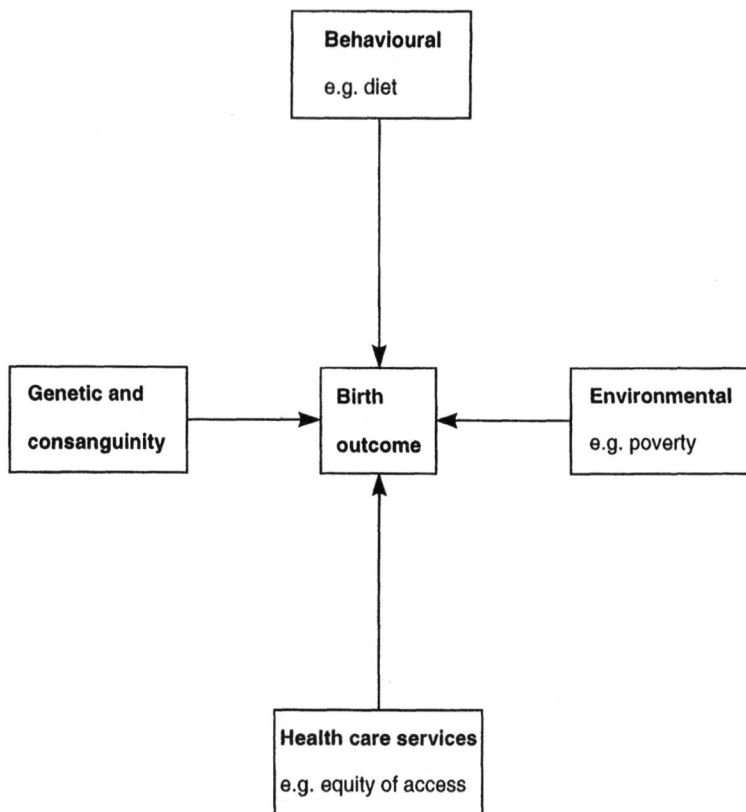

Figure 6.1   Factors which affect birth outcome.

and cardiac malformations (Balarajan et al. 1989). The latter two are also recognised as being associated with other confounding variables – see above.

However, not all reports suggest that consanguinity is a 'bad thing'. There was no effect on birth weight in Birmingham (Honeyman et al. 1987); gestational age and neonatal anthropometric measurements in Saudi Arabia (Wong & Anokute 1990); and neonatal and perinatal mortality in Peterborough, although the numbers were small and the populations had different causes of death (Dryburgh 1986).

Consanguinity does confer important social benefits on the family. These include reassurance during periods of cultural isolation after immigration; strengthening of family ties by assisting in the retention of family land and property; and continuing the family as the centre of power and prosperity (Proctor & Smith 1992). It is unlikely to diminish as it is an important part of the culture of British Pakistanis.

Social–Environmental Factors

It is now widely recognised that low birth-weight rates and associated perinatal mortality are related to social, economic and housing deprivation (NAHA 1990). Pakistani families often come from poor socio-economic backgrounds with high levels of unemployment, poor educational attainment, poor-quality housing and overcrowding (Central Statistical Office 1994, Department of Employment 1990).

Many Pakistanis live in inner-city housing which is old, damp and of poor quality. Overcrowding is a problem, with less than 30 per cent of Bangladeshi and Pakistani households having less than one room per person (Office of Population Census and Survey 1993).

All of these adverse factors are compounded by both overt and covert racial harassment and violence.

Behavioural Factors

Antenatal calorie intakes of less than 1700 kcal per day are associated with babies who are pathologically light for dates, particularly if also associated with poor maternal weight gain during pregnancy (Dryburgh 1986, Naismith 1981).

Other risk factors which are associated with adverse pregnancy outcome and occur with an increased frequency in Pakistani women include older age, higher parity (Lumb et al. 1981) and seronegativity for rubella (Miller et al. 1987).

Health Care Services

For ethnic minority women, maternity services have been described as being inappropriate, inaccessible and inadequate (NAHA 1990). Many Asian women (79 per cent in Birmingham) are also illiterate, even in their own mother tongue (Jain 1985).

Late booking at antenatal clinics and language difficulties have been implicated in the low amniocentesis rate for Asian women in both Sheffield and Birmingham (Little & Nicoll 1988). When liaison workers were introduced the antenatal attendance increased (Raphael-Leff 1991).

The above factors were considered to be important enough for a prospective study of birth outcome in Bradford to be performed. Bradford district has the largest single Pakistani-origin population in the UK. At the time of writing, almost 13 per cent of the city's 350,000 population originate in Pakistan (District and Ward Census Digest 1991). There are approximately 6000 deliveries per annum in Bradford's two maternity units, and around a quarter of these are to women originating in Pakistan (Proctor & Smith 1992).

The purpose of this study was to examine the associations between consanguinity and the other major individual and social factors known to be associated with birth outcomes in the Pakistani community.

METHODS

All births at 20 or more completed weeks of gestation to women of Pakistani origin who lived in Bradford were included. For every two deliveries to this group of women, the next birth to a white Bradford resident of UK origin was used as a control. The delivery register in each of the two Bradford maternity units was used to identify the birth to a white woman which followed the second of each pair of births to Pakistani women.

The data abstracted from the obstetric case notes included factors known to be associated with birth outcome. These were the mother's age, gravidity, alcohol and tobacco consumption, major complications of pregnancy and the social class and employment status of both parents. In addition, any consanguineous relationship between each set of parents was recorded. The results of routine blood tests, such as for the serum rubella status of the mother, were also abstracted, along with the data recorded on her labour and delivery.

The measures of birth outcome included the condition of the baby at birth, gestational age at delivery (based on the 18-week ultrasound scan), birth weight, head circumference, the presence of any overt congenital anomaly, the Apgar score at five minutes, and whether the baby was admitted to the special care baby unit.

Data collection commenced in November 1991 and was concluded 22 months later in August 1993.

RESULTS

Only the descriptive information related to birth outcome will be presented. The detailed multivariate analysis of the confounding variables will be reported elsewhere.

Patient Population (Table 6.1)

The study included 4049 women who had 4077 babies. Of these, 4035 babies were delivered in the Bradford units, 13 in the surrounding units and 28 at home. There were 28 sets of twins.

Table 6.1 Mother's ethnic origin and consanguinity (percentages given in brackets).

	Mother's ethnic group	
	Pakistani	White
First cousins	1 495 (55.3)	1 (0.1)
Distant relatives	584 (21.6)	0
Total related	2 079 (76.9)	1 (0.1)
Unrelated	622 (23.0)	1 347 (99.9)
Total	2 701 (100)	1 348 (100)

In the study population there were 2701 women of Pakistani origin, 692 of whom were born in the UK, and 1348 white women of UK origin. Within the Pakistani group, just over three quarters of the women were related to their partners, while only one white couple were related. This solitary white couple has been excluded from subsequent analysis. The majority of Pakistani consanguineous couples were recorded as being first cousins, the remainder being second cousins or more distantly related. It is virtually impossible to make an accurate estimation of the extent and complexity of consanguineous family pedigrees based solely on the data obtained at the booking interview. All births to related Pakistani couples were therefore aggregated into one group for analysis and compared to births to women of Pakistani origin who were unrelated to their partners. Data on births to white women who were unrelated to their partners were compared to all births to Pakistani women.

Table 6.2 Some specific characteristics of Bradford mothers and their partners (percentages given in brackets).

| | Pakistani | | | White |
	Related	Unrelated	Total	Total
Age				
<20	159 (7.6)	35 (5.6)	194 (7.2)	111 (8.2)
20–24	880 (42.3)	207 (33.3)	1 087 (40.2)	389 (28.8)
25–29	544 (26.2)	178 (28.6)	722 (26.7)	478 (35.5)
30–34	301 (14.5)	125 (20.1)	426 (15.8)	280 (20.8)
35+	195 (9.4)	77 (12.4)	272 (10.1)	90 (6.7)
Total	2 079 (100)	622 (100)	2 701 (100)	1 348 (100)
Gravidity				
0	598 (28.8)	158 (25.4)	756 (28.0)	400 (29.7)
1	464 (22.3)	142 (22.8)	606 (22.4)	454 (33.7)
2	329 (15.8)	97 (15.6)	426 (15.8)	224 (16.6)
3	215 (10.3)	80 (12.9)	295 (10.9)	149 (11.0)
4	473 (22.8)	145 (23.3)	618 (22.9)	121 (9.0)
Total	2 079 (100)	622 (100)	2 701 (100)	1 348 (100)
Smoking before pregnancy				
	23 (1.2)	18 (2.9)	41 (1.5)	516 (38.3)
Drinking alcohol before pregnancy				
	2 (0.1)	5 (2.9)	7 (0.2)	247 (18.3)
Partner unemployed				
	865 (42.6)	237 (39.4)	1 102 (41.9)	244 (19.9)

Risk Characteristics of Mothers and Their Partners (Table 6.2)

Age
The age distribution showed differences between Pakistani and white women. There were proportionally more Pakistani women aged 35 and over but the ages of white women were spread more evenly over the 20–34 age range than those of Pakistani women, 40.2 per cent of whom were in the 20–24 age range. Within the Pakistani group, the women who were related to their partners tended to be younger than those who were unrelated. The age distribution in this group was similar to that of white women.

Gravidity
Similar proportions of white and Pakistani women were experiencing their first pregnancy, but a much higher proportion of Pakistani

women had experienced four or more previous pregnancies. There were no differences between related and unrelated Pakistani couples in this respect. Among women who had experienced four or more pregnancies, 8.4 per cent of Pakistani women but only 1.6 per cent of white women were aged 35 years or more.

Smoking and alcohol
White women were far more likely to smoke than Pakistanis. Although only 41 Pakistani women smoked, smoking was less uncommon among those who were unrelated to their partners. The patterns of alcohol consumption were similar, with very few Pakistani women reporting drinking.

Employment status
Pakistani women were twice as likely as white women to have a partner who was unemployed. Although the proportion of unemployed partners appeared slightly higher when women were related to them, the difference was no greater than would be expected by chance.

Birth Outcomes (Tables 6.3, 6.4 and 6.5)

Late foetal losses
Twelve Pakistani women experienced late foetal losses at 20–23 weeks of gestation. No white women experienced such losses. No difference could be detected between the rates of foetal loss in the two Pakistani groups.

Stillbirths
There were 21 stillbirths to Pakistani women and eight to white women. No difference could be detected between the rates of stillbirth of white and Pakistani women or between the two Pakistani groups. As the study relies on data from the obstetric case notes, complete data about neonatal deaths were not available.

Gestational age
On average, Pakistani babies were delivered at a lower gestational age than white babies. There were also higher proportions of preterm and very preterm births to Pakistani women.

Birth weight
No differences could be found between the two Pakistani groups in either mean birth weight or rates of preterm births. White babies had a higher

mean birth weight and a lower proportion of babies weighing under 1000 grams. No differences were detected between the Pakistani groups.

Table 6.3 Number (rate*) of foetal deaths and live births (with 95 per cent confidence intervals).

| | Pakistani | | | White |
	Related	Unrelated	Total	Total
Foetal deaths at 20–23 weeks	8 (3.8)	4 (6.4)	12 (4.4)	0
95% CI	1.6–7.5	1.8–6.3	2.3–7.6	
Stillbirths at 24 weeks or more	17 (8.1)	4 (6.4)	21 (7.7)	8 (5.9)
95% CI	4.7–13.0	1.8–6.3	4.8–11.8	2.6–11.7
Livebirths	2074	620	2694	1341
Total registrable births	2091	624	2727	1349

* Rates used are:
Foetal death rate is the number of cases per 1000 total births (live and dead) at 20 or more completed weeks of gestation.
Stillbirth rate is the number of cases per 1000 total births (live and dead) at 24 or more completed weeks of gestation.

Table 6.4 Mean birth weight, head circumference and gestational ages (with 95 per cent confidence intervals) for babies born at 24 weeks or over.

| | Pakistani | | | White |
	Related	Unrelated	Total	Total
Birth weight (g)				
Mean	3085	3146	3099	3316
95% CI	3061–3109	3101–3191	3078–3121	3285–3344
N	2068	619	2687	1346
Head circumference (cm)				
Mean	33.9	34.0	33.9	34.5
95% CI	33.8–34.0	33.9–34.1	33.9–34.0	34.4–34.6
N	2039	611	2650	1339
Gestational age (weeks)				
Mean	39.0	39.0	39.0	39.5
95% CI	39–39.1	38.8–39.2	38.9–39.1	39.4–39.7
N	2079	622	2701	1347

Apgar scores
One-minute and five-minute Apgar scores were recorded. The five-minute scores are shown in Table 6.5. Compared with white babies, a higher proportion of Pakistani babies had Apgar scores of under 5 at 5 minutes. Once again, there was no difference between the two Pakistani groups.

Table 6.5 Birth outcomes (N and %) and relative risk (RR) with 95 per cent confidence intervals.

	Pakistani			White
	Related	Unrelated	Total	Total
Gestational age of 28 weeks or less	23 (1.1)	7 (1.1)	30 (1.1)	4 (0.3)
RR		0.98		3.74
95% CI		0.84–1.53		1.32–10.6
Gestational age of 36 weeks or less	185 (8.8)	49 (7.8)	234 (8.7)	77 (5.9)
RR		1.13		1.48
95% CI		0.84–1.53		1.15–1.89
Birth weight <1000g	22 (1.0)	5 (0.8)	27 (1.0)	3 (0.2)
RR		1.31		4.45
95% CI		0.5–3.45		1.35–14.6
Apgar score under 5 at 5 minutes	57 (2.7)	19 (3.0)	76 (2.8)	20 (1.5)
RR		0.90		1.89
95% CI		0.54–1.5		1.2–5.43
Severe anomaly at 24 hours	33 (1.6)	8 (1.3)	41 (1.5)	8 (0.6)
RR		1.23		2.55
95% CI		0.57–2.49		1.2–5.43

Congenital anomalies
Similar patterns were observed in the proportions of babies with severe anomalies. In particular, no difference was detected in the proportions of babies who had severe anomalies born to related and unrelated parents.

Other confounding variables
Pakistani mothers were significantly older, had a significantly higher number of previous pregnancies, were less likely to be rubella immune and had a significantly higher proportion of unemployed partners than

white women. Within the Pakistani group, however, those unrelated to their partners had a higher mean age at 27.2 years compared with 25.9 years for women related to their partners. Although the numbers were minimal, unrelated Pakistani women were also more likely to smoke and drink during their pregnancy.

DISCUSSION

In this study, we found that data on consanguinity were recorded in the case notes of each woman, regardless of ethnicity. Our anxieties about the accuracy of these data were addressed by conducting a smaller, qualitative study (Proctor et al. In preparation). The interview data from 49 couples revealed in many cases that the nature of a couple's consanguinity was more complex than the case notes implied. Generally, however, the overall gradient of the closeness of relationships in consanguineous couples was similar. Couples identified as first cousins in the case notes were usually more closely related than those identified as distant relatives or those unrelated to their partners.

At 76.9 per cent, the proportion of Pakistani couples with consanguineous marriages in the study is higher than previous estimates (Bundey et al. 1989, Baker et al. 1988). Even among the UK-born Pakistani women, 75.3 per cent of the marriages were consanguineous. As consanguinity has important social and cultural functions for the Pakistani community, its increase in popularity is probably multifactorial. In addition, it may be linked to current immigration policies (Parsons et al. 1993).

The maternal factors known to adversely affect birth outcome were more strongly associated with the mother's ethnicity than with consanguinity.

For maternal age, there are women, usually of a higher socioeconomic background, who postpone child-bearing deliberately until they are older, and others who continue to reproduce from their youth until middle age. This second group of women was at higher risk of both maternal and foetal morbidity. A much higher proportion of Pakistani mothers aged 35 and over were 'grand' multigravidae, and thus fall into this second group of high risk women.

The low rates of smoking and alcohol consumption among Pakistani women were not unexpected (Ahmad et al. 1989, Naismith 1981). Of the 41 Pakistani smokers, 66 per cent were born in the UK.

Smoking rates among Asian women are generally lower than among white women, but this may change, particularly among second and third generations. The relationship between socio-economic deprivation and birth outcome is well documented (Miller et al. 1987, Jain 1985). We chose not to apply the usual social class categorisation to the occupation of either the women or their partners in our study for two main reasons. First, where there is a history of long-term unemployment, the use of social class based on present or most recent employment has questionable validity. Second, there is evidence to suggest that within occupational categories, members of black minority groups tend to occupy lower positions and have lower incomes, and are more likely to work unsociable hours. There were stark differences between the employment status of Pakistani and white couples. The adverse effects of unemployment may be further compounded by health problems associated with living in poor-quality, overcrowded inner-city housing (Little & Nicoll 1988). Inner-city residence is associated with poorer quality health care, including during pregnancy (Raphael-Leff 1991, District and Ward Census Digest 1991). Recent figures suggest that families originating in Pakistan or Bangladesh are twice as likely than any other ethnic group to live in the 'worst quality' housing (Proctor et al. In preparation).

No association was found between consanguinity and stillbirth, birth weight, head circumference, gestational age, low Apgar score at 5 minutes or severe anomalies apparent by 24 hours after delivery. However, babies whose parents were related had more congenital anomalies apparent after 24 hours (chi-square = 3.94; DG (degrees of freedom) = 1; $p < 0.05$).

The congenital anomalies were those which were identified up to 24 hours after delivery. This is a problem, as some cardiac and metabolic conditions are often not diagnosed until much later after the birth (Parsons et al. 1993). Therefore, our data relating to congenital anomalies should be viewed with some caution.

Compared with white babies, more Pakistani babies were of low birth weight, were preterm, had low Apgar scores or had severe congenital anomalies.

The validity of the Apgar score as an assessment of the condition of new born babies is a subject of debate. However, it is still used routinely in many units as an early assessment. The higher proportion of Pakistani babies with Apgar scores of 5 or less at 5 minutes is a matter of concern. One possible reason for this is the high level of subjectivity in allocating the Apgar score points relating to babies' colour.

CONCLUDING COMMENTS

The differences between white and Pakistani women in the factors which can affect birth outcomes and in the outcome for the babies in the study were much wider than the differences between related and unrelated Pakistanis.

Birth outcome in Bradford's Pakistani community is associated with a number of complex and interrelated factors (Whitehead 1992). These include the level of genetic risk as expressed by the coefficient of inbreeding, which itself varies with the degree of consanguinity (Proctor et al. In preparation); social class and associated deprivation; institutional racism; and epidemiological and clinical obsession with consanguinity as the principle cause of this problem.

Consanguinity is a common practice amongst this community. It is known to increase the risk of certain inherited genetic disorders, but the findings of this study suggest that it is not the main factor associated with adverse birth outcome in Bradford Pakistani families.

ACKNOWLEDGEMENTS

We would like to thank the Obstetric, Midwifery and Clerical Staff at the Bradford Maternity Units for their support and cooperation throughout the duration of the study.

REFERENCES

Ahmad, W.I.U., Kernohan, E.E.M., & Baker, M.R. 1989 Health of British Asians: a research review. *Community Medicine* 11: 49–56.
Baker, M.R., Kernohan, E.E.M., Bevan, P., & Knight, T. 1988 The contribution of consanguinity to the poor outcome of pregnancy in Asians in Bradford. Paper presented to the annual meeting of the Society for Social Medicine, 14–16 September.
Balarajan, R., Soni Raleigh, V., & Botting, B. 1989 Mortality from congenital malformations in England and Wales: variations by mothers' country of birth. *Archives of Diseases of Childhood* 64: 1457–62.
Bundey, S., & Alam, H. 1993 A five year prospective study of the health of children in different ethnic groups, with particular reference to the effect of inbreeding. *European Journal of Human Genetics* 1: 206–19.

Bundey, S., Alam, H., Kaur, A., Mir, S., & Lancashire, R.J. 1989 Race, consanguinity and social features in Birmingham babies: a basis for prospective study. *Journal of Epidemiology and Community Health* 44: 130–5.
Central Statistical Office 1994 *Social Trends – 24*. HMSO: London.
Chitty, L.S., & Winter, R.M. 1989 Perinatal mortality in different ethnic groups. *Archives of Diseases of Childhood* 64: 1036–41.
Darr, A., & Modell, B. 1988 The frequency of consanguineous marriage among British Pakistanis. *Journal of Medical Genetics* 25: 186–90.
Department of Employment 1990 Ethnic origins and the labour market. *Employment Gazette* 98: 125–37. HMSO: London.
District and Ward Census Digest 1991. Research Section, City of Bradford Metropolitan Council: Bradford.
Dryburgh, E.H. 1986 Neonatal problems in the Asian population of Peterborough. *Midwife, Health Visitor and Community Nurse* 22: 27–30.
Gillies, D.R.N., Lealman, G.T., Lumb, K.M., & Congdon, P. 1984 Analysis of ethnic influence on stillbirths and infant mortality in Bradford, 1975–81. *Journal of Epidemiology and Community Health* 38: 214–17.
Honeyman, M.M., Bahl, L., Marshall, T. et al. 1987 Consanguinity and fetal growth in Pakistani Moslems. *Archives of Diseases of Childhood* 62: 231–5.
Jain, C. 1985 *Attitudes of Pregnant Asian Women to Antenatal Care*. West Midlands Regional Health Authority: Birmingham.
Little, J., & Nicoll, A. 1988 The epidemiology and service implications of congenital and constitutional anomalies in ethnic minorities in the UK. *Paediatric and Perinatal Epidemiology* 2: 161–84.
Lumb, K.M., Congdon, P.J., & Lealman, G.T. 1981. A comparative review of Asian and British born maternity patients in Bradford, 1974–78. *Journal of Epidemiology and Community Health* 35: 106–9.
Miller, E., Nicoll, A., Rousseau, S.A., Sequiera, P.J.L., Hambling, N.H., Smithells, R.W., & Holzel, H. 1987 Congenital rubella in babies of South Asian women in England and Wales: an excess and its causes. *BMJ* 295: 734–9.
Naismith, D.J. 1981 Diet in pregnancy: recommendations and realities, in Bateman, E. C. (ed.) *Applied Nutrition 1*. John Libbey: New York.
National Association of Health Authorities (NAHA) 1990 Review of services for black and ethnic minority people. *Maternity Services Bulletin* no. 2. NAHA: Birmingham.
Office of Population Census and Survey 1993 *Mortality Statistics 1991. Perinatal and Infant: Social and biological factors England and Wales* 25. HMSO: London.
Parsons, L., Macfarlane, A., & Golding, J. 1993 Pregnancy, birth and maternity care, in Ahmad, W.I.U. (ed.) *Race and Health in Contemporary Britain*. Open University Press: Milton Keynes.
Proctor, S.R., Karbani, G., Mueller, R.F., Howlett B.C., Macfarlane, A., Ahmad W.I.U., Marfell, K., & Smith, I.J. In preparation. Factors associated with birth outcome in Bradford Pakistanis.
Proctor, S.R., & Smith, I.J. 1992 A reconsideration of the factors affecting birth outcome in Pakistani Muslim families in Britain. *Midwifery* 8: 76–81.
Raphael-Leff, J. 1991 Ethnic and cultural aspects of maternity care. *Maternal and Child Health* 16: 145–6.

Terry, P.B., Condie, R.G., & Settatree, R.S. 1980 Analysis of ethnic differences in perinatal statistics. *BMJ* 281: 1307–8.

Whitehead, M. 1992 *The Health Divide*, 2nd edition. Penguin Books: London.

Wong, S., & Anokute, C.C. 1990 The effect of consanguinity on pregnancy outcome in Saudi Arabia. *Journal of the Royal Society of Health* 110: 146–7.

Young, I. 1990 Hereditary disorders, in McAvoy, B.R. & Donaldson, L.J. (eds) *Health Care for Asians*. Oxford University Press: Oxford.

7 The Relevance of Cultural Understanding to Clinical Genetic Practice

Nadeem Qureshi

Two decades ago, physicians in the USA made the unfortunate mistake of trying to provide mass sickle cell screening without counselling (Whitten 1973). Fortunately, with time, we came to appreciate the importance of appropriate counselling. Our understanding has now evolved further, and consideration of the cultural background of the individuals we are approaching is quite rightly considered an important issue. In the next few pages this issue will be explored and its relevance to community-based clinical genetics practice will be considered.

GENETICS SERVICES FOR MINORITY GROUPS

In providing genetic information to families from minority groups, there are two tiers to the service. First, there are the services required for families who have children affected by specific genetic disorders. Second, there are the genetic screening and counselling services for the rest of the community. The introduction of haemoglobinopathy services to the North-East London Cypriot population demonstrates an appropriate organisation of the latter (Mouzouras et al. 1980). This approach, however, cannot simply be applied without modification to other minority groups because of the different circumstances of each community.

In organising these two tiers of genetics services, some general points are worthy of mention. An understanding of the cultural background of the community being approached is vital to ensure that the appropriate genetic information does reach those in these minority groups who would like to receive it. Cultural understanding is not only an issue for minorities, however, but needs to be considered in developing services for the whole population. Furthermore, culture is not a

111

static phenomenon but is constantly changing. For example, the contemporary British communities of Southern Asians have different expectations from their communities of origin in Asia; they also now differ from the first-generation communities in Britain that have been intensively researched during the last two decades. This makes ongoing research into the changing cultural expectations of minority communities particularly relevant.

Community and Individual

The cultural norms of a minority population can be considered at various levels. The community's views, as depicted by 'community leaders', may be quite different from those of individual families. For instance, on a community level, termination of pregnancy may be discouraged, whilst in a confidential counselling session, a family from this community may feel more comfortable with this approach (Bundey et al. 1991). The utilisation of prenatal diagnosis by the Bradford Muslim community is another practical example. Mid-trimester foetal diagnosis of thalassaemia major had little uptake in this community, but this has increased with the introduction of first-trimester foetal testing by chorionic villus sampling (Modell et al. 1984). Even within families, attitudes can vary between generations and between individuals.

In order to disseminate genetic information effectively to minority groups, it is important to understand their cultural background (Wang & Marsh 1992). Within this context, an appreciation of the dynamics of the family is important. This helps to assess how (i.e. through which family members) genetic information may most appropriately be channelled. In some circumstances, as when approaching the parents of affected children, there may not be a choice. In other settings, however, such as in a community genetic screening programme, there may be a choice of which family members to approach. In this case, it may be most appropriate to approach the individuals who hold influential and supportive roles within the family. In the British Asian communities, in the not too distant past, this would have been the eldest member of the household (Glendenning 1979). As the elders have lost their control of the purse-strings, however, this position of social influence has gradually been eroded. In the 'modern' British Asian family, young married couples may be seen as the most appropriate channel for disseminating genetic information. These family members can provide an 'informed nucleus' within the extended family, facilitating the propagation of genetic information to their relatives. Such approaches are not restricted to minority groups, as demonstrated by Krush and Evans (1984).

Other influential persons within the family would be the wife of the eldest brother and the eldest sister. They are often seen by younger married women in the family as sources of domestic and social advice. An appropriate term for such a person could be the 'prominent matriarch' (although some would object to such a phrase, considering the concept of matriarch as synonymous with the wife's mother-in-law, and hence a dictatorial role). All these routes for transmitting information should be utilised by the community genetics services.

Community Clinical Genetics Nurse or Genetics Associate

In order to discover the most appropriate potential 'informed nuclei' in the extended family, one needs to employ a community-based minority health care worker, ideally a community clinical genetics nurse, but other appropriately trained primary health care workers may provide a more cost-effective approach. This individual needs to be from a similar cultural and religious background to the community for whom services are being developed. For instance, a Mirpuri Punjabi would be likely to trust a Mirpuri health worker, provided the latter displayed a confidential and professional manner. Considering that the information provided would be disseminated through the women in the extended family, the health care worker should also be female. At present, the recruitment of such ethnic minority workers is difficult because there are few women with the appropriate qualifications. An integral part of any post would therefore need to be an in-service training programme (Dixson et al. 1992).

The primary health care team, including family doctors, health visitors and receptionists, could also give valuable insights into the extended families registered in their practices and the appropriate contacts in these families.

Health Education in Genetics

For an effective campaign, one needs appropriate information provided in the media that are read, listened to and watched by the minority group. This needs extremely sensitive packaging to avoid generating stigmatisation and anxiety. There is a danger in any health education campaign that the targeted group may feel that a finger of blame is being pointed at them. Furthermore, the health education campaign needs to provide the individuals with the sense that they 'own' the information and that they control how it is used. They must feel that they want the information and that it is not being imposed on them. The mandatory screening programme for sickle cell that was implemented in the USA in the 1970s is an extreme example of such imposition of unsought information (Whitten 1973).

Bearing these cautionary notes in mind, any attempt to disseminate genetic information to or within minority groups needs to be closely observed and fully evaluated as a pilot scheme before an all-out campaign is launched. It is interesting that there is already a mass of research into the psychological effects of cystic fibrosis carrier screening, which has only been possible since 1989. In contrast, in the mid-1980s such research was not seriously considered for the haemoglobinopathy carrier screening programme (Johnson et al. 1993). At the same time, minority communities are becoming increasingly apprehensive about research focusing only on their specific communities. To avoid such problems, any future research needs to have clearly stated objectives, and to be of potential long-term benefit to the target community. Perhaps, in the future, research ethics committees will assess these concerns before granting ethical approval.

CONSANGUINITY: FAMILY COHESION, PROFESSIONAL BLAME

Health professionals' ignorance of the structure and dynamics of consanguineous marriages can result in a communication breakdown between health care workers and members of extended families from ethnic minority groups. This marital pattern is practised by many British Muslim communities. The Muslim communities in Britain amount to nearly two million people, the largest being the British Pakistanis. The majority of British Pakistani marriages are still arranged, and it is still common for one of the partners to come from Pakistan.

In considering the impact of genetic morbidity on a community, one has to look beyond medical issues by also considering the socio-cultural and psychological issues. For example, one would find that at the time of a stressful event, such as the birth of a 'chronically ill child', the family may respond differently depending on the marital pattern of the parents of the affected child. Figure 7.1 demonstrates

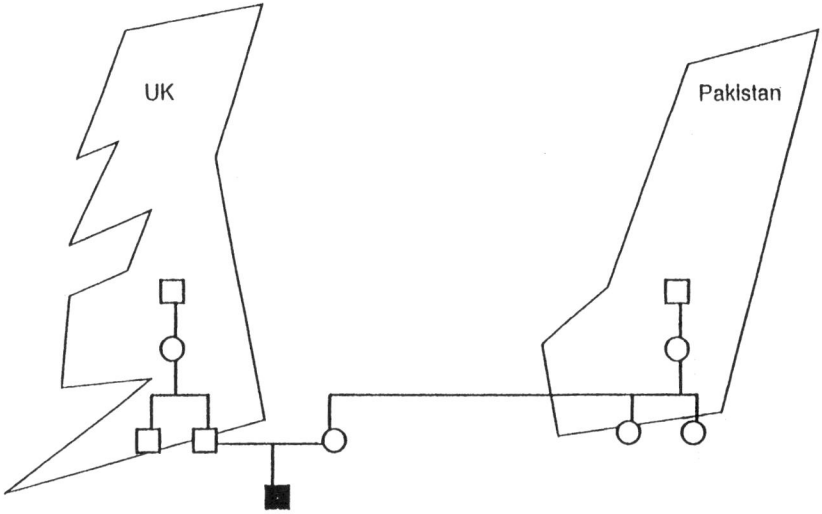

Figure 7.1 Family networks in a non-consanguineous marriage with a child affected by genetic disease.

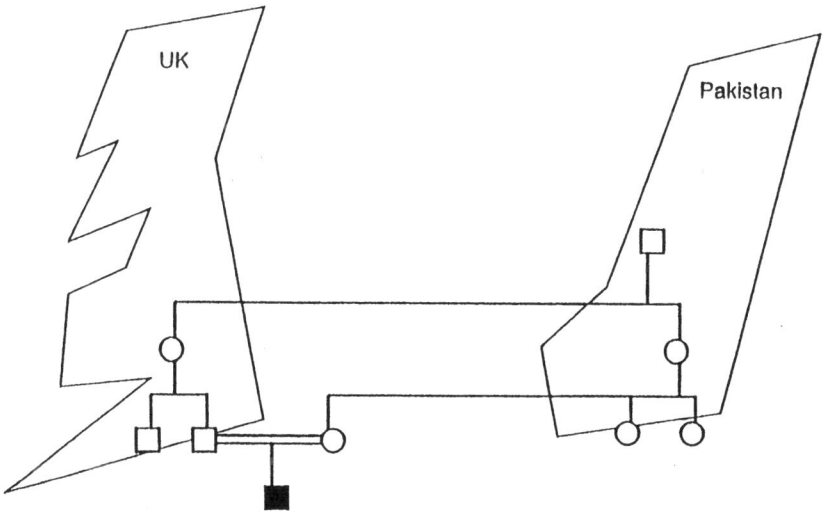

Figure 7.2 Family networks in a consanguineous marriage with a child affected by genetic disease.

the possible scenario in a non-consanguineous marriage. It is not uncommon for the 'outsider', in this case the wife, to be blamed for the whole family's tragedy. Having been brought into a new country and a new family, she is trapped with little support in an alien environment. Figure 7.2 illustrates the same medical problem in a different family context, where the parents are cousins. Darr and Modell (1988) have demonstrated that the proportion of marriages in the British Pakistani population which are consanguineous is increasing. If we consider the birth of an affected child in such a marriage, we see that – despite being away from her parents and siblings – the wife can still turn for support to her aunt, who is also her mother-in-law (Khlat et al. 1986). The socially stabilising role of this marital pattern is often overlooked by the medical establishment (Darr 1990). Consanguinity can help the extended family to act as a single cohesive unit.

This support network, however, can be disrupted by inappropriate statements pronounced by medical practitioners or others: for instance, 'Your child is abnormal because you are cousins.' Such statements not only produce strain within the couple, but also disrupt the whole extended family. It is like driving a 'medical nail' through the whole family. Such harsh, judgemental statements run totally counter to the principles of genetic counselling that have been adopted in clinical practice. Just as one would never consider blaming a non-immunised mother for her child's congenital rubella, so one should never blame a couple for their child's illness; to do so is potentially most destructive.

Some would argue that directiveness is more appropriate in minority groups (Wang & Marsh 1992), but such comments simply reinforce the prejudices already held by many health professionals. This does not negate the worth, however, of providing accurate risk information, before conception, to young consanguineous couples. By providing such information, all couples should be helped to make informed reproductive decisions.

Another display of the poor understanding by health professionals of consanguinity and marital patterns is illustrated by the comment 'We were surprised to find that Mr Khan and Mrs Shah, both with similarly affected children, were related.' This demonstrates that genetic counselling cannot be restricted to the 'at risk' couple, but this information has to be disseminated throughout the extended family (Figure 7.3).

Implications for Genetics Services

Many health professionals feel apprehensive about discussing such delicate issues as consanguinity with the extended family, fearing

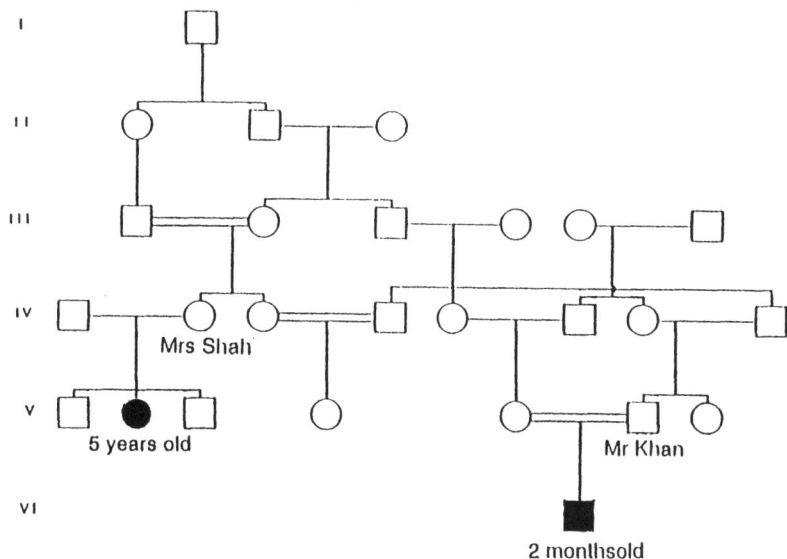

Figure 7.3 Extended family network illustrating how the same genetic disorder can arise in two separate, but related, nuclear families.

that they may create feelings of stigmatisation. This can be avoided, however, if proper services are provided.

The clinical genetics services should employ appropriately trained British Pakistanis as genetic counsellors (Modell & Kuliev 1992). As previously discussed, these counsellors need to be from a similar cultural and social background to the local community, in this case the British Pakistani community. In this way they will have the knowledge and sensitivity to tackle difficult situations. Also, stigmatisation is reduced within a closely related extended family through the appropriate counselling of certain family members (the 'informed nuclei').

In an attempt to draw these issues together within a practical public health model, let us consider how minority groups may be approached in the community by a clinical genetics service.

Within any community or population, a certain proportion of individuals will develop genetic disorders. The medical profession will already be aware of some of the families with affected members ('conspicuous genetic morbidity'). These details are often held in general practice and hospital medical records, and the information is likely to be disseminated within the particular families through the affected individuals or their parents. The other group to be considered are those families with affected members who are not known, at present, to the medical profession ('hidden genetic morbidity'). These families could be discovered by a programme of accurate family history recording, population genetic screening to identify carriers of recessive disorders (such as the haemoglobinopathies) and diagnostic screening procedures. Considering diagnostic screening procedures, Bundey et al. (1991) have suggested that high-resolution antenatal ultrasound examination in consanguineous marriages will identify 70 per cent of lethal malformations.

The level of hidden and conspicuous genetic morbidity in all communities still needs investigating. The prevalence and severity of genetic morbidity in a community is important in planning future community genetics services. The findings of Bundey and Alam (1993) with respect to the level and nature of genetic morbidity in the Birmingham Pakistani community, for example, will help in planning local services in the Birmingham area.

The response of a community to screening for a genetic disease is highly dependent on the cultural norms of that population and the cultural dynamics within each family. Insight into the family structure enables the clinical genetics services to provide counselling and education through appropriate 'informed nuclei' within each family. This can be facilitated by relevant health education campaigns and media presentations.

The cultural expectations of the various minority – *and majority* – groups in this country are not dramatically different, and we need to be cautious about overemphasising these. However, when tackling sensitive issues like genetic screening and counselling, the approaches that are used do need fine tuning to be relevant to the families who are being offered the service. This can be facilitated by community genetic counsellors who have been recruited from the relevant ethnic group and have the appropriate training.

REFERENCES

Bundey, S., & Alam, H. 1993 A five year prospective study of the health of children in different ethnic groups, with particular reference to the effect of inbreeding. *European Journal of Human Genetics* 1: 206–19.
Bundey, S., Alam, H., Kaur, A., Mir, S., & Lancashire, R.J. 1991 Why do UK born Pakistani babies have high perinatal and neonatal mortality rates? *Paediatric and Perinatal Epidemiology* 5: 101–14.

Darr, A. 1990 The social implications of thalassaemia among Muslims of Pakistani origin in England: family experience and service delivery. Ph.D. thesis, University of London.

Darr, A., & Modell, B. 1988 The frequency of consanguineous marriage among British Pakistanis. *Journal of Medical Genetics* 25: 186–90.

Dixson, B., Dang, V., Cleveland, J.O., & Peterson, R.M. 1992 An educational program to overcome language and cultural barriers to genetic services. *Journal of Genetic Counselling* 1: 267–74.

Glendenning, F. 1979 *The Elders in Ethnic Minorities*. Beth Johnson Foundation Publications (in association with the Commission for Racial Equality): London, pp. 50–3.

Johnson, A.G. (chair) 1993 *Report on a Working Party of the Standing Medical Advisory Committee on Sickle Cell, Thalassaemia and other Haemoglobinopathies*. HMSO: London, section 3.

Khlat, M., Halabi, S., Khudr, A., Der Kaloustian, V.M. 1986 Perception of consanguineous marriages and their genetic effects among a sample of couples from Beirut. *American Journal of Medical Genetics* 25: 299–306.

Krush, J.K., & Evans, K.A. 1984 *Family Studies in Genetic Disorder*. Charles C. Thomas: Springfield, Ill.: chapter 5.

Modell, B., & Kuliev, A.M. 1992 Social and genetic implications of customary consanguineous marriage among British Pakistanis. *Galton Institute Occasional Papers*, second series, no. 4.

Modell, B., Petrou, M., Ward, R.H.T., Fairweather, D.V.I., Rodeck, C., Varnavides, L.A., & White, J.M. 1984 Effect of fetal diagnostic testing on the birth-rate of thalassaemia in Britain. *Lancet* II: 1383–6.

Mouzouras, M., Camba, L., Ioannou, P., & Modell, B. 1980 Thalassaemia as a model of recessive disease in the community. *Lancet* II: 574–8.

Wang, V., & Marsh, F. 1992 Ethical principles and cultural integrity in health care delivery: Asian ethnocultural perspectives in genetic service. *Journal of Genetic Counselling* 1: 81–92.

Whitten, C.F. 1973 Sickle-cell programming: an imperiled promise. *New England Journal of Medicine* 288: 318–19.

8 Women's Experiences of Screening in Pregnancy: Ethnic Differences in the West Midlands

Josephine Green and Merry France-Dawson

INTRODUCTION

Screening for foetal abnormalities has become a major part of antenatal care. Virtually all women in the UK will have at least one ultrasound scan during their pregnancy, as well as blood tests to screen for certain conditions in the mother herself or in her baby.

Our study arose from a need to discover more about the attitudes and experiences of pregnant women with regard to this screening. How much do they know about the tests that they have? Are they reassured by them, or made anxious? How do anxieties about foetal abnormality fit in with other sources of anxiety? To what extent are different reactions to the screening process a result of differences in hospital procedures (at one extreme) or personality differences (at the other)?

Reviewing the previous literature, it is clear that screening and diagnostic tests during pregnancy do have the potential to be both 'calming' and 'harming' (Green 1990). Our aim was to go beyond the relatively narrow focus of most previous studies and to examine women's experiences of routine screening within the context of other aspects of their lives in order to answer some of the questions raised above.

The study had two parts. The main study was based on nine hospitals all within a 50-mile radius of Cambridge and had a predominantly Caucasian sample. The second part of the study (the sickle cell study) was based on two hospitals in the West Midlands and was particularly focused on ethnic differences and on Afro-Caribbean women's experiences of screening for sickle cell conditions. These two hospitals were selected for the racial diversity of people they served. This paper reports on two particular areas of interest in the sickle cell study: women's access to prenatal screening and their knowledge and experience of sickle cell screening. Some of the main study findings have been published (and are listed at the end of this chapter) and other papers are in preparation.

METHODOLOGY AND RESULTS

Data were collected by means of questionnaires that women completed at booking and at approximately 22 and 35 weeks antenatally and six weeks postnatally. Women of African descent in the sickle cell study received an additional questionnaire at 22 weeks about sickle cell screening. The questionnaires were designed specifically to chart women's experiences and feelings about the things that happened to them during the course of the study; they contained a mixture of standardised measures and narrative-style questions. All women booking for antenatal care who could read and write English were eligible for the study.

Women in the sickle cell study were recruited in the antenatal clinic at the time of their booking visit. Thereafter, all questionnaires were posted directly to their homes. The sampling strategy was to recruit 200 unselected women from each hospital, and thereafter to restrict the sample to women of African extraction to ensure sufficient numbers for whom sickle cell testing was potentially an issue. The numbers of African women arriving for their first hospital visit was, in fact, much lower than expected, and the initial period of data collection (three months) was extended for a further three months. These additional women were recruited by the midwife.

A total of 538 women were recruited. Demographic characteristics are given in Table 8.1. Thirty-one per cent of the women were primiparous.

Thirty-one women who reported that they were positive for the sickle cell gene (i.e. they were either carriers or they had a sickle cell condition) were subsequently also interviewed in their homes.

Timing of First Hospital Appointment

Women of African descent in the sample were significantly more likely to arrive for their first hospital appointment later than other women: only 57 per cent were booked in by 16 weeks gestation, compared to 73 per cent of Asian and 75 per cent of Caucasian women (Figure 8.1).

Table 8.1 Sample characteristics (N=538).

Age mean = 26.2 sd = 5.4

Ethnicity	N	%
African	288	53
Asian	44	8
Caucasian	206	38
Total	538	100

Education		
No schooling	1	0
Up to 16 yrs old	287	53
16–18 yrs old	167	31
Over 18 yrs old	37	7
Still at school	6	1
Not given	40	7
Total	538	100

Religion		
None	69	13
Protestant/CoE	276	51
Catholic	40	7
Christian sects	47	9
Muslim	9	2
Hindu	30	6
Buddhist	1	<1
Other	6	1
No, inappropriate answer	60	11
Total	538	100

Late booking clearly restricts women's options with regard to screening and diagnostic tests during pregnancy, and we therefore felt that it was important to investigate the reasons for this.

The first questionnaire had asked women how many weeks pregnant they had been when they first saw their GP regarding the current pregnancy. The modal response was eight weeks. There were no differences between the ethnic groups. Thus, the later hospital booking of women of African descent cannot be ascribed to later initial contact with the GP.

A short questionnaire was sent to 12 GPs who referred patients to the two study hospitals, asking about the screening services they provided for black women. Only four (two from each district) replied. These stated that a standard procedure was being followed for the referral of all women for antenatal care, irrespective of ethnicity.

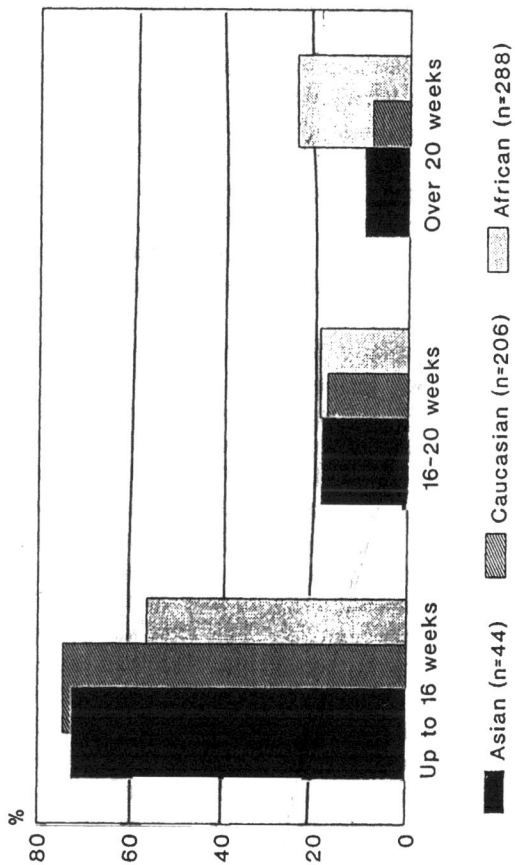

Figure 8.1   Weeks pregnant at hospital booking (sickle cell study).

Midwives at the hospital seem to believe that although African descent women appear to be arriving late, they do not appear to miss first appointments any more than any other group. Indeed, at the time of data collection, when the researcher observed that African women seemed to be more heavily pregnant at booking, some letters and appointments were checked to confirm that this was the first appointment given.

Seven GPs were interviewed regarding the screening services they provided, to see whether their referral procedure for women of African ancestry was in any way compromised. Five of the GPs were from an Asian background. The two Caucasian GPs did not seem to think that there were any major problems regarding getting women to the hospital for early booking, though one of them thought that 'black women need to be persuaded'. Three of the Asian GPs indicated that Asian women enjoyed going to the hospital, as it was a break from the routine of their homes. Asian women, they thought, particularly liked being in hospital for childbirth, as for many the break seemed like a holiday and the family made a great fuss of them. When asked about women of African ancestry, four of the Asian GPs indicated that first-time mothers are always keen to get to the hospital early for their tests, but experienced mothers preferred not to attend until it was absolutely necessary. As one GP put it, 'Many of my [sic] black mothers come from countries where there is little or no maternity care. They are used to having their children naturally, so they do not want anybody doing anything to them. They can cope.' This view was not widely expressed by the women who were interviewed. Although most said that they preferred to have as natural a pregnancy and childbirth as possible, only one women said that she actively avoided going to hospital until the last possible minute (including for delivery). This woman had had an unpleasant hospital experience with the birth of her first child and actively contrived to avoid hospitals whenever she became pregnant. Two of her five children were born at home; a third was born in the ambulance en route to the hospital, another just arriving at the hospital in time for its birth.

Another four women admitted to 'not liking hospitals' though they tried to attend when requested. The rest of the women preferred to have tests done early enough 'so that they know what's what'. As one woman said, 'My kids will have a hard enough time going through life without me bringing out sickly ones. If they're fit, they've got a chance. They can fight back.'

Familiarity with Names of Specific Conditions

Women were presented with a list of conditions: thalassaemia, sickle cell anaemia, Tay-Sachs disease, Down's syndrome, spina bifida, haemophilia, muscular dystrophy, cystic fibrosis, anencephaly and cerebral palsy. They were asked which they had heard of (see Figure 8.2). Most had not heard of anencephaly (92 per cent), Tay-Sachs (90 per cent) and thalassaemia (85 per cent). The majority of women of African descent had heard of sickle cell, compared with just under half the Asians and two thirds of the Caucasians. Caucasian women were significantly less likely to have heard of thalassaemia than African and Asian women. Asian women were significantly less likely to have heard of nearly all other conditions on the list.

Sickle Cell Testing

Women of African descent were sent an additional questionnaire specifically about sickle cell testing; 159 were returned.

Fifty-two per cent of those who returned the questionnaire (N=83) had been tested for sickle cell at some point in their lives, 37 per cent (N=58) said they had never had a sickle cell test, while another 11 per cent (N=18) said that they were not sure.

Of the women who had not been tested for sickle cell or were not sure (N=76), 31 (41 per cent) wanted to be tested. Only seven women (10 per cent of those not tested) did not want the test, but 35 women (46 per cent) said that they did not know whether they wanted to be tested, mainly because they were concerned about what the test entailed.

Twenty-eight women had been tested before ever becoming pregnant and 65 of the 83 had been tested during either a previous pregnancy or the current one. Forty-five women, 28 per cent of the sample, were tested during this pregnancy, 42 of whom gave information about timing. Only 52 per cent (N=22) of these were tested at 16 weeks gestation or less. A further 19 per cent (N=8) were tested by the 20th week, with 29 per cent (N=12) being tested after the 20th week.

Thirty-six per cent of those tested said that they had clear explanations of the sickle cell test (Table 8.2). Of those who responded in the 'other' category, 3 per cent had prior knowledge of the test, with one woman saying that she obtained some information from the Health Education Authority's *Pregnancy Book.* Four per cent said that they only realised they had been tested when they received their results in

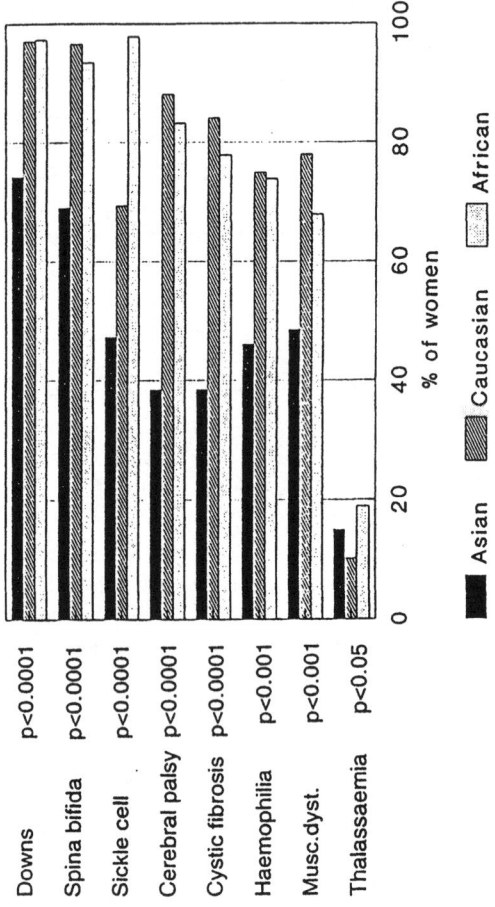

Figure 8.2 Women who had heard of various conditions, by ethnic group.

the post. A further 4 per cent simply said that they were given no information at all.

Women were asked a number of other questions about their experiences of sickle cell testing. Fifty-four per cent had been worried to some degree about having the test. There was an overwhelming view (90 per cent of the sample) that the results of the test should always be given: only 9 per cent thought that results need only be given if they were positive. Two thirds of those tested were given a result, although only 25 (30 per cent of those tested) said that they had it in writing.

Table 8.2 Responses to: 'Before having the test, did you feel that you knew and understood as much about it as you needed to?' (N=83)

	N	%
Yes, everything was explained very clearly	30	36
I think they told me a lot of things but I didn't really take it in	12	14
I would have liked to know more but didn't like to ask	13	16
I didn't realise there was anything to know	6	7
I asked for information but didn't get any	3	4
I didn't want to know very much	2	2
Other (please say what)	14	17
Missing	3	4

Terminology

One of the difficulties revealed by other work in this area is one of terminology (France-Dawson 1990). A 'trace of sickle cell', instead of 'sickle cell trait', is a simplistic expression commonly used by both the women and their doctors. This is often explained as only a small amount of sickle cell gene in the blood, i.e. not enough to cause illness. Women who had been given results were asked what they understood by this term, and also by 'being a carrier' and 'having sickle cell'. A 'trace of sickle cell' was explained by 13 per cent as meaning that the gene could be passed on to their children, and a further 23 per cent said that it meant that an individual was a 'carrier' but gave no indication that they knew what this meant. Thirty-eight per cent (N=20) believed that the individual had a serious illness (Figure 8.3). The term 'carrier' was explained by 50 per cent of those answering the question in terms of the possibility of a child inheriting the gene (N=23), although some people still believed that the individual was

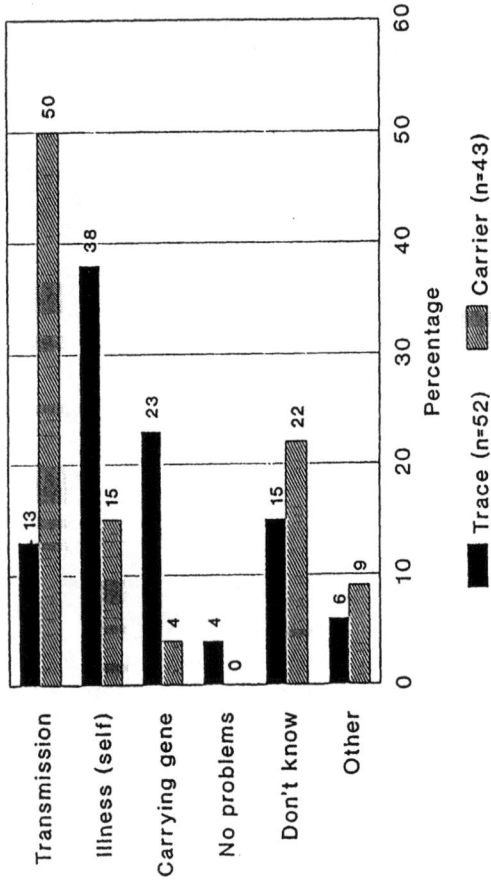

Figure 8.3 Responses to: 'If told you have a trace / are a carrier of sickle cell, what do you understand?'

at risk from varying degrees of illness (15 per cent, N=7). One person thought that they could pass the sickle cell anaemia to their siblings.

Of the 41 women who answered the question about what it meant to be told they had sickle cell (an expression often used), 54 per cent (N=22) spoke in terms of having a serious illness, and 17 per cent said they didn't know what it meant. The remaining respondents gave a variety of answers ranging from transmission to siblings and childbirth difficulties to 'the baby would degenerate' and 'positive for the gene'.

DISCUSSION

In this sample, women of African descent were not gaining access to maternity care services as early as other women. The reasons for this are still not entirely clear, as there are differences of opinion between the small numbers of doctors seen and the women interviewed. Furthermore, the women who were interviewed may not be representative because their sickle cell status may have made them more willing to have early maternity care, although some of them were only made aware of their status during this pregnancy.

The GPs attributed later arrival for hospital booking to African women's aversion to medical intervention. It is possible that older, fertile, immigrant women do prefer as little medical intervention as possible, especially if they have already had one or more relatively trouble-free pregnancies. However, the average age of the women in this sample makes it highly likely that most were British born. Certainly the majority of those interviewed were born in Britain. The few who were born elsewhere emigrated here when they were small children. Whilst their parents' cultural beliefs may have influenced how they perceive medical care, they would be very much 'between cultures', accepting the notion of antenatal care as the norm. Whatever the explanation for late booking in this particular sample, an interesting issue is raised about the perceptions held by one ethnic (and professional) minority group – Asian GPs – of another ethnic (and gender) minority group – women of African descent. This issue deserves further study.

Our sample was selected primarily to compare women of African descent with Caucasian women. The inclusion of some Asian women was incidental, and the Asian sample is not large enough to investigate all the questions of interest. However, despite this, it is quite clear that Asian women knew little about disorders for which testing might be offered, and this lack of information shows up in other aspects of the data which we have not reported here. The relative ignorance of the Asian women in this sample is surprising, given their fluency in English and their level of education, and raises particular concerns about other Asian women to whom these circumstances do not apply.

Although most women of African descent had heard of 'sickle cell', they in fact knew little about the conditions themselves. They also appeared to have received very little information about the tests and the sickle cell conditions during their pregnancies. The data show that more people associate having a 'trace of sickle cell' with having an illness than they do if the word 'carrier' is used. More importantly, those who were told they had a 'trace' of sickle cell seemed to be less aware of the implications for transmitting the gene. A few women, who recognised that there would be few or no problems in terms of their health, also seemed to be unaware of the implications for transmission of sickle cell anaemia when the male partner is also a carrier.

Using the expression 'sickle cell' for 'sickle cell anaemia' may also bring about problems. Approximately 80 per cent of the sample spoke of sickle cell as being a major problem. It therefore causes problems when they receive letters which simply say 'you have been found to be positive for the sickle cell gene'. One woman said how lucky she had been so far because she had 'sickle cell but did not get sick like some' she knew. She worried constantly about when it would 'come to cause her problems' as she hoped 'it wouldn't take her' before her child was old enough to fend for himself. On checking her haemoglobinopathy card, she was found to be a carrier.

Poor information provision before and after testing and the use of ambiguous terms when information is given appear to be adding to the confused picture black women have about sickle cell conditions. Test results sent by post should be accompanied by the appropriate information leaflets which include contact addresses and an invitation to discuss the results with a named health professional.

It seems clear that sickle cell screening is not occurring on a major scale in this particular population. When it occurs, it is often too late for fully informed decision-making on the part of the women involved. Nearly half of those who were tested during this pregnancy had been tested after 16 weeks; indeed, nearly half of the women of African descent were more than 16 weeks at booking. This clearly compromises their access to other routine tests (e.g. serum screening for

Down's syndrome) and limits their options in the event of a positive result. The data from this study therefore suggest that pregnant women of African descent in the West Midlands are not receiving an optimal service.

ACKNOWLEDGEMENTS

This research was funded by the Health Promotion Research Trust.

REFERENCES

France-Dawson, M. 1990 *Sickle Cell Conditions – the Continuing Need for Comprehensive Health Care Services: A study of patients' views.* Daphne Heald Research Unit, RCN: London.
Green, J.M. 1990 Calming or harming? A critical review of psychological effects of fetal diagnosis on pregnant women. *Galton Institute Occasional Papers*, second series, no. 2.

RELATED PUBLICATIONS

Green, J.M. 1992 Principles and practicalities of carrier screening: attitudes of recent parents. *Journal of Medical Genetics* 29(5): 313–19.
Green, J.M., & Murray, D. 1994 The use of the EPDS in research to explore the relationship between antenatal and postnatal dysphoria, in Cox, J.L., & Holden, J. (eds) *Prevention of Depression after Childbirth: Use and misuse of the Edinburgh Postnatal Depression Scale.* Gaskell Press: London.
Green, J.M., Snowdon, C., & Statham, H. 1993 Pregnant women's attitudes to abortion and prenatal screening. *Journal of Reproductive and Infant Psychology* 11: 31–9.
Green, J.M., Statham, H., & Snowdon, C. 1991 EPDS by post. *British Journal of Psychiatry* 158: 865.
Green, J.M., Statham, H., & Snowdon, C. 1992 Screening for fetal abnormalities: attitudes and experiences, in Chard, T., & Richards, M.P.M. (eds) *Obstetrics in the 1990s: Current controversies.* Mac Keith Press: London.
Green, J.M., Statham, H., & Snowdon, C. 1993 Women's knowledge of prenatal screening tests. 1: Relationships with hospital screening policy and demographic factors. *Journal of Reproductive and Infant Psychology* 11: 11–20.

Green, J.M., Statham, H., & Snowdon, C. In preparation. *Pregnancy: A testing time.* Report of the Cambridge Prenatal Screening Study. Centre for Family Research, University of Cambridge: Cambridge.

Richards, M.P.M., & Green, J.M. 1993 Attitudes toward prenatal screening for fetal abnormality and detection of carriers of genetic disease: a discussion paper. *Journal of Reproductive and Infant Psychology* 11: 49–56.

Statham, H., & Green, J.M. 1994 The effects of miscarriage and other 'unsuccessful' pregnancies on feelings early in a subsequent pregnancy. *Journal of Reproductive and Infant Psychology* 12: 45–54.

Statham, H., Green, J., & Snowdon, C. 1991 When is a fetus a dead baby? *Lancet* 337: 856.

Statham, H., Green, J., & Snowdon, C. 1992 Psychological and social aspects of screening for fetal abnormality during routine antenatal care. *Proceedings of 'Research and the Midwife',* November 1992.

Statham, H., Green, J., & Snowdon, C. 1993 Mothers' consent to screening newborn babies for disease. *British Medical Journal* 306: 858–9.

Statham, H., Green, J., Snowdon, C., & France-Dawson, M. 1993 Choice of baby's sex. *Lancet* 341: 564–5.

9 Coping with the Sickle Cell Gene in Africa

Olu Akinyanju

The gene that concerns us most in Nigeria is the sickle cell (S) gene, which causes sickle cell anaemia (SCA), also known as sickle cell disease (SCD). SCA is the most common hereditary disorder in the world (WHO 1994), and is caused by the inheritance of a mutant sickle haemoglobin (HbS) from each parent. Its incidence is highest in sub-Saharan Africa (Livingstone 1967), from where the gene has spread by migration to Europe and especially the Americas. To the title of this volume – Culture, Kinship and Genes – I would like to add Geography, as it is the geographical endowment of sub-Saharan Africa – hot, lush and humid – which has favoured the survival and growth of the gene among the indigenous population.

In the intervening stages the mosquito thrived, and still does, in this environment; the mosquitoes, through their bites, then transmitted the deadly *Plasmodium falciparum* malaria parasite from one person to the other. Many of the people who were thus infected with malaria, especially infants and young children, died as a result; but many of those who carried the sickle cell gene as heterozygotes (Hb AS) did not die – they survived to reproduce and pass the gene on to some of their children (Allison 1954, Edington & Watson-Williams 1965, Livingstone 1971). As these heterozygous carriers of the sickle cell trait were healthy, possession of the trait thus represents a natural adaption for human survival in an otherwise hostile environment. On the other hand, those who had SCA by virtue of being sickle gene homozygotes (Hb SS) readily succumbed to malaria, leading to the loss each time of two S genes. The majority who were homozygous for the usual adult haemoglobin type (Hb AA) survived worse than the AS but better than the SS.

Thus, as far as survival in the environment was concerned, the sickle gene homozygotes (SS) were unfit, the unaffected majority with haemoglobin A genes (AA) were fitter and the sickle gene heterozygotes (AS) were fittest. Gradually, generation after generation over the centuries, the survival of the fittest led to the enlargement within the population of the S gene pool. The sickle cell gene has also thrived in other areas of the world where the falciparum malaria has been or is currently prevalent. These include countries in the Mediterranean basin and South India (Livingstone 1967).

Conversely, a study of population genetics has shown that the eradication of falciparum malaria, by eliminating the reproductive advantage of Hb AS heterozygotes, predictably leads to a gradual contraction of the S gene pool within a population. Hence, in South Africa and southern Mozambique, both of which lie within the temperate non-malarial zone of sub-Saharan Africa, the S gene frequency is so low that sickle cell anaemia is not perceived there as a problem. Heterozygotes (Hb AS) are less than 0.5 per cent of the Bantu population in these areas (Bernstein 1969, Reys et al. 1972), in contrast to the much higher prevalence among Bantus in northern Mozambique and in countries north of the Zambesi River. The 2000-year-old migration of Bantus from West to South Africa is conjectured to have led to this reduction of the S gene pool there. The same argument is advanced for the reduction in the S gene frequencies in African–Americans and African–West Indians. Thus, the ultimate control of the S gene within a population is linked to the eradication of malaria in that population.

The problem in the affected countries in Africa is that malaria is not being controlled; Luzzatto (1975) has estimated that, even if this were instantly achieved, it would still take some 300 years for the gene frequency to be reduced by half. As a strategy for reducing the incidence of SCA, it is too slow to be appealing to countries where sickle cell disorders constitute a significant public health problem. Undoubtedly, the immediate beneficiaries from control of malaria would be the affected homozygotes with Hb SS, whose survival through childhood would be better assured (Molineaux et al. 1979).

In the context of sickle cell, any reference to sub-Saharan Africa should exclude the areas south of the Zambesi River where the absence of lethal malaria has not favoured the sickle gene. We should correctly refer to the affected region as tropical sub-Saharan Africa. In this area, four haplotypes denoting four independent origins of the sickle gene have been discovered, namely the Senegalese, the Benin, the Bantu and, most lately, the Cameroon haplotypes (Pagnier et al. 1984, Trabuchet et al. 1991). The recent discovery of the Cameroon haplotype, with its circumscribed distribution, suggests that more haplotypes are almost certainly waiting to be discovered in tropical Africa.

The size of the clinical problem in tropical sub-Saharan Africa is enormous – almost inconceivable. Estimates of the prevalence of heterozygote carriers (Hb AS) vary from 14 to 45 per cent in the communities of various countries (Konotey-Ahulu 1991). In Nigeria, several studies (Jellife & Humphreys 1952, Walters & Lehmann 1956, Fleming et al. 1979) and a recent unpublished national survey have shown the prevalence of AS heterozygotes in the adult population to be about 25 per cent. In Africa, there are estimated to be more than 240 000 births of affected infants (Hb SS) every year, of which about 90 000 are in Nigeria (Akinyanju 1989).

It is strange that, despite the high incidence of sickle cell disorders, many communities in Africa have scarcely been aware of the problem until recently. This is because the underlying genetic diagnosis of SCA in the many affected dead infants has been unrecognised – it has been submerged within the larger pool of non-sickle-related infant mortality. This is not to say that the disorder was not recognised at all. It was, judging from the numerous descriptive indigenous names given to the disorder. The fact is that sickle cell anaemia was self-limited to the paediatric age group. It has only been in recent times, following the introduction of effective anti-malarial and other anti-infective measures, that the affected individuals broke through the paediatric age survival barrier to be recognised in adolescence and adulthood. In the 1950s, SCD was clearly better recognised in Africans of the diaspora, such as African–Americans, than in indigenous Africans. Hence, in a 1950 review of sickle cell disease in Africa and America, Raper concluded that, 'This essay has been directed to showing that the disease is of more importance to the American negro than to the African'. To explain this, he suggested an admixture of African and Caucasian genes as responsible for the higher incidence of sickle cell anaemia in African–Americans. This is, of course, completely wrong. What Raper had failed to realise was that the higher prevalence of SCA in African–Americans was not an indication of a higher incidence in them than in native Africans. In tropical Africa, there was always a higher incidence of SCA at birth but a lower prevalence in the population owing to higher infant mortality rates.

More than 40 years after Herrick's first report of a patient with SCA in America (1910) and six years after Raper's report, Mabayoje – a Nigerian physician working in Lagos – wrote:

SCA is a major disease of West Africa. It is a cause of distress in many families. It therefore deserves much better recognition than it gets at present. It should be treated as a major disease in schools of tropical medicine, textbooks, and in all medical schools in tropical Africa. (Mabayoje 1956)

Sadly, his observations remain as true today as they were when he made them. The present situation is that no one now doubts the preponderance of SCA in Africa. With increasing control of communicable diseases, chronic and hereditary ones, such as SCA, have now assumed a position of major public health concern in each of the 34 countries in the affected region of Africa. To quote Molineaux et al. (1979):

There is no other known inherited disorder present at such high frequency in a large population and of comparable severity as sickle cell anaemia in Africa. With rising standards of living and control of malaria, sickle cell anaemia will become an immense medical, social and economic problem throughout the continent.

Factors necessary in addressing the problems created by sickle cell in Africa include peace, political will and resources – probably in that order of importance. In most of the countries involved, none of the above can be demonstrated. Many of the countries, having been arbitrarily created in the 19th and 20th centuries by colonial powers, without regard for ethnic boundaries or cultural affinity, are presently undergoing the pains of violently redefining themselves or reordering their leadership or political structure. Without peace, political will and resources will be wanting. Even in peaceful countries, the emergence of the necessary political will cannot be taken for granted. Our experience in Nigeria is that political will has to be coaxed and nurtured by the effective sensitisation and education of policy-makers and resource-allocators.

SICKLE CELL DISEASE AS A PUBLIC HEALTH PROBLEM

A few years ago, we were informed by WHO officials that the issue of control of SCA had not been presented as a major public health concern by the African Ministers of Health at their annual World Health Assembly meeting with the WHO Director General in Geneva. There are several possible reasons for this glaring omission. First, the countries involved have a myriad of problems competing for attention.

Second, gains previously made in the control of communicable diseases are sometimes not sustained, for one reason or another, and the resurgence of these diseases again compels priority attention. Third, there is much conflict in Africa, and even seemingly peaceful countries are often politically unstable and suffer from a high turnover of ministers and other key officials, with consequent disruption to the process of developing priorities and strategies. Fourth, there is no regional professional group dedicated to the evolution and implementation of appropriate strategies for addressing the problems associated with sickle cell disorders in Africa. Such a group would need to be capable of systematically sensitising, educating and advising the policy-makers in each country, and would be committed to the implementation of policy.

The last of the three factors mentioned earlier which are required to address the issue of malaria – resources – is very scarce. The poorest countries in the world are in Africa, and some of them have recently been described as suffering from 'an absolute resource lack' (Godlee 1994). Since the mid-1980s, massive devaluation of currencies and galloping inflation have mortally wounded the economies of many countries in the region. To make matters worse, no international aid agency has adopted sickle cell disorders as a problem worthy of support. Even the WHO tried unsuccessfully to persuade wealthy international donor agencies to help finance a control programme recommended by an expert panel it had convened in early 1987 (WHO 1987).

Late in 1987, at a WHO training course in Greece, I was privileged to meet two other African doctors involved with the care of patients with sickle cell disorders in the Republics of Cameroon and Zambia. The three of us were so impressed with the progress being made with thalassaemia in Europe by the WHO Working Group on the Community Control of Hereditary Anaemias, and the contrasting lack of activity on sickle cell in Africa, that we wrote a joint letter to the WHO Regional Director for Africa. We urged him to establish an African Working Group on Sickle Cell Disorders. Possibly as a result of our letter, the Regional Director invited to his office in Brazzaville, Congo, in October 1991, an eight-member 'Regional Study Group on Haemoglobinopathies' to advise him on pertinent preparation for a meeting, scheduled for 1992, of national health authorities on the public health implications of haemoglobinopathies on the African continent. The task was accomplished but the larger meeting was never convened, allegedly because of lack of funds.

With the serious problems outlined above, it is clear that it would be a long time before any effective government or, sadly, even WHO-promoted initiative on sickle cell in Africa is seen. The absence of any coherent programme of control in the face of a growing adult population of affected individuals will surely be disastrous. As nature abhors a vacuum, the glaring absence of appropriate services will create a field day for charlatans, stunt the development of research, encourage misinformation and stigmatisation and generally increase the misery in affected families.

THE SICKLE CELL CLUB

Surely the time was ripe for private initiative. This is what we thought in 1984 when we founded the Sickle Cell Club Lagos, Nigeria. We have fortunately had a measure of success, and I hope that an account of our activities and experiences since then will encourage others with similar problems to chart a path of fruitful private initiative in their own countries. Second, I hope that this account will convince some international donor agencies, including the WHO, that SCA is of great public health importance in Africa and is deserving of their serious attention.

The original Sickle Cell Club of Nigeria was formed in Ibadan in 1972 with the objectives listed in Table 9.1. Branches were established in all towns with university teaching hospitals; it was chiefly haematologists or paediatricians who were appointed as principal officers. The club made very little impact nationally and branches thrived in only one or two places where the leadership had demonstrated drive and resourcefulness. Once these leaders left town (one of them being Scottish) the branches promptly collapsed. This was the situation in 1984 when we decided to revive the Lagos Branch as an autonomous Sickle Cell Club Lagos, Nigeria. The only possible hitch was smoothed over when we persuaded an enthusiastic television producer, who herself had sickle cell anaemia, that it was unnecessary to form a brand new patient support association as it might well lead to unnecessary rivalry with attendant duplication of effort and confusion.

The failure of the original Sickle Cell Club of Nigeria is discussed here as a case study of interest to anyone who would like to set up a patient/family support association. Our observations led us to believe that the club failed principally for three reasons. First, it was too centralised in structure for an association that had no administrative organ for guiding, coordinating or monitoring the activities of its branches in such a large country. Second, leaders, especially of the branch clubs, were selected more for their academic or clinical attainments than for their managerial skills, drive or resourcefulness. It was wrongly assumed that any haematologist or paediatrician, given his or her knowledge of the condition, could run a Sickle Cell Club. Third, lay members of the public, including the affected individuals and their families for whom the clubs were formed, were not integrated into the leadership structure of the clubs. Thus, the leadership unintentionally assumed a patronising mien.

Table 9.1 Objectives of the Sickle Cell Club of Lagos, Nigeria.

The objectives of the Club are:

1. To ensure the establishment and proper functioning of Sickle Cell Clinics throughout Lagos, Nigeria.
2. To publish simple booklets for patients, relatives and the general public to educate them about the disease and its management.
3. To issue identity cards to patients and members affected so that they can be treated without delay whenever admitted to any hospital.
4. To give assistance in education to patients who have missed schooling through sickness.
5. To help to secure suitable employment for the handicapped.
6. To provide genetic counselling.
7. To arrange social activities where members can meet and discuss problems related to the disease and their possible solutions.
8. To arrange social activities for the purpose of fund raising.
9. To support research aimed at finding means of alleviation of sickle cell disease and related disorders.
10. To help in the training of laboratory staff and equipping hospital laboratories in simple methods of diagnosing sickle cell disease.
11. To publish periodic news-sheets for doctors and others in which advances in the treatment of the disease are discussed.

In order to avoid repeating the failed recipe, from the very start the revived Sickle Cell Club had only two doctors on a nine-member executive committee. We decided to create a secretariat with paid staff in order to have at our disposal the administrative and executive capacity to facilitate and implement decisions. As Nigeria is a large country with poorly developed intercity communication, we favoured the development of strong local clubs and planned to focus on the problem in Lagos while stimulating people from other areas to establish Sickle Cell Clubs in their own localities. The Sickle Cell Clubs are registered as non-profit-making non-governmental organisations (NGOs). We recognised the limitation in resources and therefore, rather than attempt to achieve all the objectives listed in Table 9.1, we selected a few that could be met and that would have the most ripple effect. We decided that training was a key factor in empowering members and health workers to embark successfully on new club formation, advocacy, counselling or treatment.

By 1985, only one Nigerian, a paediatrician, had undergone training in counselling on sickle cell disorders. Incidentally, this was at the Brent Sickle Cell Centre, London. As we needed to train many Nigerians, we decided that it would be more cost-effective to invite the trainers from London to Lagos for the training in counselling, rather than attempt to send the trainees to London (Akinyanju & Anionwu 1989). The five annual courses so far held have proven very popular and have always been oversubscribed. A total of 150 Nigerians and four professionals from other African countries have been trained. In the process, a few of those trained earlier have themselves served as competent trainers.

Other courses held have included an update seminar for medical laboratory technicians on the 'Laboratory diagnosis of haemoglobin disorders'; a seminar on 'Understanding sickle cell' for journalists and other public media personnel; and three annual update seminars on 'Current concepts in the management of sickle cell disorders' for doctors and nurses.

Apart from these educational and training activities, we have participated in public enlightenment programmes in newspapers and on radio and television, and we have delivered lectures to school children and to other members of the public on sickle cell disorders. One week in the year is designated 'Sickle Cell Week', during which the attention of the public is particularly focused on SCA through well-publicised programmes and through the organisation of fund-raising bazaars and dinners. Funding for our programmes has come through these activities and from direct appeals to individuals or corporations. Donation of services and materials are frequently sought and have helped tremendously in achieving specific objectives. The enrolment of corporate members, who are able to pay enhanced annual subscriptions and contribute various skills when necessary, has also been very useful.

We have succeeded in sensitising government personnel, mainly officials of the Ministry of Health, to the issues involved in the management of sickle cell disorders by inviting them as special guests to the opening or closing ceremonies of our programmes and special events. In this way, we have kept them informed of our activities, and we have been assured of their approval.

Of all our activities, easily the most rewarding has been the training in counselling. Over the years, we have deliberately given preference to i) participants with a nursing background, and ii) those sponsored by their employers; members of these groups are the most likely to be given the opportunity to put their newly acquired skills into practice. The training has empowered counsellors to act as advocates on behalf of patients with SCD and to organise and lead new Sickle Cell Clubs in their own localities. In this way, new Sickle Cell Clubs have been formed so far in ten other places in Nigeria, and these have been linked by common membership of an established Federation of Sickle Cell Clubs of Nigeria. Individuals who have successfully undergone the training enrol as members of the Sickle Cell Counsellors Association of Nigeria, which is dedicated to the development of culturally sensitive counselling skills and to campaigning for the establishment of effective services.

APPROACH TO GENETIC COUNSELLING

There is now evidence that the introduction of counselling in Nigeria has been beneficial to the patients and their families. The affected patient load is so great that it constitutes the bulk of the clientele for counselling. In order to cope with large numbers in one medical centre, a counselling strategy has been devised in which counselling is delivered to age peer groups who are expected to have similar problems, while individual counselling is reserved as a follow-up for clients who appear to require it.

Individual counselling remains the norm where the counsellor/client ratio permits. Its impact was recently evaluated in a centre in Lagos, where it was found that school and job absences were significantly reduced after counselling. Furthermore, the counselled patients were all much better informed on inheritance of SCD and demonstrated greater self-confidence and self-esteem (Table 9.2). The counselled parents of affected children demonstrated – or at least admitted to – less despair, less anxiety, less scepticism and less hostility to their children (Tables 9.3 and 9.4). Thus, even before the introduction of other specialised services, counselling of affected individuals and their parents has proven to be beneficial in the management of patients and families with sickle cell disorders in Nigeria.

Table 9.2 Effect of counselling 500 SCD patients.

	Pre-counselling	2 years post-counselling
Correct information on inheritance of SCD	10%	100%
Confidence and self-esteem	25%	95%

Based on unpublished observations of Otaigbe and Akinyanju.

Table 9.3 Feelings of parents before and after counselling 500 SCD families.

	Initial session	2 years later
Hostility and/or despair	80%	2%

Table 9.4 Attitude of parents to affected children before and after counselling 500 SCD families.

| | Before counselling | | After counselling | |
	Mothers	Fathers	Mothers	Fathers
Over-protective	85%	60%	10%	2%
Sceptical	15%	30%	NIL	2%
Rejecting	NIL	10%	NIL	NIL

A Lagos clinic survey of 263 mothers of patients with SCA showed that 97 per cent of them wanted a prenatal diagnostic (PND) service to be introduced; 84 per cent preferred the chorionic villus sampling in early pregnancy over the sampling of amniotic fluid in later pregnancy. Although it is not expected to reduce significantly the incidence of SCA in Nigeria, the introduction of PND was in great demand as an option in making reproductive choices. Prenatal diagnosis was therefore introduced in 1993 by inviting to Lagos two London-based obstetricians versed in the art. They succeeded in transferring the necessary skills to two Nigerian obstetricians. Although access to the procedure is restricted by cost, it is far better than it was and can be further improved over time. The skills can also be transferred to other centres in Nigeria and in neighbouring countries with a view to increasing its coverage and accessibility in the sub-region.

Our programmes have required commitment, resourcefulness, goodwill and sometimes sheer luck to succeed. We are not, however, deceived into believing that we have solved any problems except that of showing that private initiative is a viable alternative where lack of resources and low government prioritisation limit the services that can be provided. In retrospect, the self-help private initiative route appears to have a number of advantages in the evolution of effective strategies and programmes for addressing the problems of sickle cell in Africa. First, money is not wasted as it is not available anyway. Second, self-help tends to compel more efficient use of resources and an optimal search for and utilisation of non-monetary contributions. Third, it is easier to identify committed, altruistic workers because those with mercenary motives are unlikely to be attracted to the struggle. Fourth, it circumvents bureaucratic delays commonly associated with national public programmes.

Before I sound as if I am extolling poverty, I should point out that lack of funds has worsened the plight of many affected families, and will continue to do so until it is reversed. The discipline imposed by poverty is no substitute for the coverage that adequate funds would permit. A middle way – a happy mean – should be found. What is recommended is that money should not simply be thrown at the problem, but should be applied to discovering how best to deliver appropriate services that can be integrated into the existing health care systems of the countries in sub-Saharan Africa.

CONCLUSION

Given the inertia and the very limited resources in Africa, how then should one kick-start such programmes in African countries? A worthwhile method would be to identify a core of committed health care professionals and officials in each country, and to provide them with training in counselling and in other aspects of the management and control of sickle cell disorders, as well as in the organisation of support groups and services for the affected population. The importance of such a project should be obvious as it should stimulate and empower potential leaders in the field with a sound basis of relevant knowledge and skills. Such a project should also prevent or limit the emergence of well-meaning but naive efforts, which may prove to be counter-productive. One of the lessons quickly learned by us in Nigeria is that the creation of awareness should not be an end in itself but a means to an end – that end being appropriate services. One should always resist the temptation to launch hopefully into indiscriminate awareness campaigns unless backed by the capacity to provide appropriate education and at least some essential services.

One recurring struggle that has to be fought is against the desire of some people to advocate overt or subtle coercion in the choice of 'the right' marital partners as an effective strategy for reducing the incidence of SCD. Fortunately, this struggle can be won by educating the opposition: these are people who are aware of the mode of inheritance of SCD but have not considered the problems likely to be caused by this policy. Of course, no such reduction in the incidence of hereditary disorders has ever resulted from such a policy. On the contrary, it would introduce suspicion, concealment of sickle trait status and unhelpful stigmatisation, as was the experience in Cyprus (Angastiniotis & Hadjimanas 1981).

In most of tropical Africa, the perception of SCD is of a painful, costly and lethal affliction. Even the most privileged families can lose affected members rapidly at any age. This contrasts with the growing view among American and European workers of sickle cell anaemia as a relatively mild condition in which a fatal outcome can almost always be averted by appropriate care. This latter viewpoint is often advanced in defence of the position that, unlike patients with thalassaemia major, those with sickle cell anaemia should not, on demand, be given potentially curative, though radical, treatment, such as bone marrow transplantation. Although some have attributed the latter position to muted racism, there is also no doubt that the existing wide difference in the availability of appropriate health and welfare services between the developed and the developing countries is largely responsible for the divergence in their perception of the severity of the sickle cell syndrome.

Finally, the study of sickle cell has already aided both our general understanding of the molecular basis of disease and the development of the technology of genetic diagnosis. It is my impression that confronting the challenge of sickle cell disorders in Africa will similarly enhance our understanding of effective social, community and genetic intervention in other hereditary disorders as well.

ACKNOWLEDGEMENTS

This chapter is a revised version of a paper originally given to the Culture, Kinship and Genes meeting in Abergavenny in 1994 as the Medical Research Council Guest Presentation. I am very grateful to the Medical Research Council for their financial support, enabling me to attend that meeting.

REFERENCES

Akinyanju, O.O. 1989 A profile of sickle cell disease in Nigeria. *Annals of the New York Academy of Sciences* 565: 126–36.
Akinyanju, O.O., & Anionwu, E.N. 1989 Training of counsellors on sickle cell disorders in Africa. *Lancet* 1: 653–4.
Allison, A.C. 1954 Protection afforded by sickle cell trait against subtertian malarial infection. *BMJ* 1: 290–4.
Angastiniotis, M.A., & Hadjimanas, M.G. 1981 Prevention of thalassaemia in Cyprus. *Lancet* 1: 369–71.
Bernstein, R.E. 1969 Sickle haemoglobin in South Africa. *South African Medical Journal* 43: 1455–6.
Edington, G.M., & Watson-Williams, E.J. 1965 Sickling, haemoglobin C, glucose-6-phosphate dehydrogenase deficiency and malaria in Western Nigeria, in Jonxis, J.H.P. (ed.) *Abnormal Haemoglobins in Africa*. Blackwell: Oxford, pp. 393–401.
Fleming, A.F., Storey, J., Molineaux, L., Iroko, E.A., & Attai, E.D.E. 1979 Abnormal haemoglobins in the Sudan savanna of Nigeria. I: Prevalence of haemoglobins and relationships between sickle cell trait, malaria and survival. *Annals of Tropical Medicine and Parasitology* 73: 161–72.
Foy, H. et al. 1952 The distribution of sickle cell trait and the incidence of sickle cell anaemia in the Negro tribes of Portuguese East Africa. *East African Medical Journal* 29: 247–51.
Godlee, F. 1994 The World Health Organization: the regions – too much power, too little effect. *BMJ* 309: 1566–70.
Herrick, J.B. 1910 Peculiar elongated and sickle-shaped red blood corpuscles in a case of severe anaemia. *Archives of Internal Medicine* 6: 517–21.
Jellife, D.B., & Humphreys, J. 1952 The sickle cell trait in Western Nigeria. *BMJ* 1: 405–6.
Konotey-Ahulu, F.I.D. 1991 *The Sickle Cell Disease Patient*. Macmillan: London.
Livingstone, F.B. 1967 *Abnormal Haemoglobins in Human Populations: A summary and interpretation*. Aldine: Chicago.
Livingstone, F.B. 1971 Malaria and human polymorphisms. *Annual Review of Genetics* 5: 33–64.
Luzzatto, L. 1975 Genetic counselling in haemoglobinopathies. *Dokita* 7: 65–8.
Mabayoje, J.O. 1956 Sickle cell anaemia: a major disease in West Africa. *BMJ* 1: 194–6.

Molineaux, L., Fleming, A.F., Cornille-Brogger, R., Kagan, I., & Storey, J. 1979 Abnormal haemoglobins in the Sudan savanna of Nigeria. III: Malaria, immunoglobulins and antimalarial antibodies in sickle cell disease. *Annals of Tropical Medicine and Parasitology* 73: 301–10.

Pagnier, J., Mears, J.G., Dunda-Belkhodja, O., Schaffer-Rego, K.E., Beldjord, C., Nagel, R.L., & Labie, D. 1984 Evidence for the multicentric origin of the sickle cell hemoglobin gene in Africa. *Proceedings of the National Academy of Sciences of the USA* 81: 1771–3.

Raper, A.B. 1950 Sickle cell disease in Africa and America – comparison. *Journal of Tropical Medicine* 53: 49–53.

Reys, L., Manso, C., Stamatoyannopoulos, G. et al. 1972 Genetic studies on South-Eastern Bantu of Mozambique. II: Serum groups, hemoglobins and red cell isozyme phenotypes. *Hymangenetik* 16: 227–33.

Trabuchet, G., Elion, J., Baudot, G., Pagnier, J., Bouhass, R., Nigon, V.M., Labie, D., & Krishnamoorthy, R. 1991 Origin and spread of beta-globin gene mutations in India, Africa and Mediterranea: analysis of 5 flanking and intragenic sequences of beta S and beta C genes. *Human Biology* 63: 241–52.

Walters, J.H., & Lehmann, H. 1956 Distribution of the S and C haemoglobin variants in two Nigerian communities. *Transactions of the Royal Society of Tropical Medicine and Hygiene* 50: 204–8.

World Health Organization 1987 *Proposal for a Feasibility Study on the Control of Sickle Cell Disease in Africa*. Report of a WHO Informal Consultation. WHO/HDP/SCD/87.3.

World Health Organization 1994 *Guidelines for the Control of Haemoglobin Disorders*. WHO/HDP/HB/GL/94.1

10 Cultural Influences on the Perception of Genetic Disorders in the Black Population of Southern Africa

Jennifer Kromberg and Trefor Jenkins

A multitude of rapid changes are occurring at present in South Africa but cultural transformation has been taking place for centuries, particularly since the 1400s when European explorers first set foot on the southern tip of the African continent. If culture is defined as a total way of life of a people, or the social legacy the individual acquires from the group, or that part of the environment that is the creation of man (Kluckhohn 1965), then it is obvious that the life of a group cannot remain unaltered when it comes into contact with another group, especially when the latter group has skills unknown and attractive to the former. However, when the latter group starts offering health services to the former, who, despite adopting some of the ways of the new culture, still have many of the old ways of their own culture ingrained in them, many problems may arise. It is therefore essential, if these problems are to be anticipated or forestalled, that the cultural influences on the perceptions of both genetic disorders and genetic service provision be understood.

We have been providing genetic counselling services and conducting research on genetic disorders in Johannesburg for 23 years (e.g. Jenkins et al. 1973, Kromberg & Berkowitz 1986, Jenkins 1990). Our clients have been drawn from all population groups and our research has involved a great variety of people, ranging from the San ('Bushmen') to the many different Bantu-speaking people, Asiatic Indians, Jews and Europeans, as well as those of mixed descent. The aim of this paper is to document some of the cultural issues which have arisen during the course of our work with patients from the black population, and to assess how these might influence their perception of genetic disorders and in what way they should be taken into

147

consideration when planning and providing relevant genetic services in a culturally appropriate manner. The providers can then be trained to avoid both ethnocentric attitudes and the transfer of cultural expectations, to the benefit of both counsellor–client communication and the transfer of knowledge in the genetic counselling process.

SYSTEMS OF THOUGHT

The traditional world view and the systems of thought in the black population appear to favour collective rather than individual thinking (Hammond-Tooke 1989), and this system may be accompanied by an external locus of control. As social psychologists might express it, the social schema of the body tends to predominate over the individual schema. A desirable balance between the individual and sociological schemata of the body is mandatory for competence and self-steering behaviour (Manganyi 1973). Without such a balance, slow decision-making processes may occur, and there may be a reluctance and inability in some individuals to make decisions or to take action without consulting someone else.

A recent practical example was provided by a young rural woman who, after childbirth, developed paralysis of the lower limbs and was unable to walk at all. One of the present writers (Kromberg) was asked by her fieldworkers to transport the woman to hospital, but when offered the lift she said she could not accept it (even though this meant she would have to be carried to hospital) because she had not consulted her mother. The outcome of the case was that she went to the traditional healer (who was probably more accessible than the hospital), responded to his treatments and could walk again shortly thereafter. Such lack of ability to make decisions is in itself paralysing in many instances, and can lead to the lack of uptake of genetics services, such as prenatal diagnosis. Accessibility is, however, another intertwined factor, and if a patient hears of prenatal diagnosis and then goes home to discuss it with her mother, she may never return, especially if she lives far from the facilities. Our studies show that the mother is usually consulted, wisdom often being the prerogative of the elderly, and afterwards the father of the foetus may or may not be consulted (Mofokeng 1991).

A second philosophy related to this apparently dominant (though probably changing) world view is associated with fatalism. One of the major components of culture is the complex of beliefs surrounding the relation of man to nature, man to his fellow men, and man to the supernatural powers which are believed to control the universe (Read 1966). If this control is believed to be absolute, then the individual cannot alter what happens to him or her. This belief leads patients in our genetic counselling clinic to say: 'If I have a handicapped child in utero, I am meant to have such a child, therefore I cannot do anything about it' (i.e. I cannot accept the offer of prenatal diagnosis and selective abortion).

Recently we have found an apparently high incidence of untreated epilepsy in a small sample in one Transvaal rural area, and the caregivers stated in a resigned way that they had to look after the affected children as they were (i.e. without anti-epileptic drugs). The reasons for such an attitude remain to be explored, and whether they will be compliant with treatment remains to be seen.

In line with group decision-making practices and fatalistic beliefs, is a diffuse body-image boundary differentiation. One of our studies on albinism (Manganyi et al. 1974) showed that black control subjects had a significantly more diffuse body-image boundary than did albino subjects, matched by age and sex, although both groups had diffuse body-image boundaries associated with difficulty in differentiating themselves from the group. Socio-cultural issues tend to nurture the development of diffuse body boundary characteristics and these are related to many varied areas of personality functioning. Further investigation is required if the implications for genetic counselling are to be understood. Lastly, integral to the system of thought presented in the black population is a somewhat undefined or indistinct line between life and death. This is associated with ancestor worship and the possibility for the dead to interfere with the affairs of the living, which affects both the perception of genetic disorders and ideas about causation.

DISEASE CAUSATION

Ancestor worship is surrounded by traditions and rituals which have to be followed in the correct way. Transgressions of the rules can evoke the ire of the ancestors, which can lead to all sorts of problems, including congenital abnormalities in the offspring of the transgressors. On the other hand, pleasing the ancestors can produce a happy outcome to, say, the pregnancy. One of the albino children known to us is called Gobela, meaning 'spirit of the ancestors'. Her mother, a traditional healer, believed that the spirits returned via her child to help her, and her practice prospered after the birth. She believed the child to be a good luck symbol. Other mothers have described their distress at being punished by the ancestors and therefore producing an albino child. The ancestors have also been blamed by some mothers of children with Down's syndrome for damaging the child. For the answer to the question 'why did this happen to me?' mothers of handicapped children will often consult the traditional healer or diviner, rather than the Western medical expert, who may be seen as useful for treatment but not very successful at answering the question 'why?'.

Other cultural beliefs surrounding disease causation include maternal impression; one common form it takes is: 'if you laugh at an albino when you are pregnant you will have an albino child' (Kromberg & Jenkins 1984). The breaking of strict taboos (for example, having intercourse too soon after the birth of a baby) can also lead to congenital defects or intellectual disability in the next baby (Schapera 1940). A similar situation has been described in the Navaho Indians in the USA, who ascribe illness to the breaking of taboos which guide behaviour, to contact with the ghosts of the dead, or to the malevolence of someone who has resorted to witchcraft (Read 1966).

COMMON PRACTICES AND TABOOS IN THE PAST AND PRESENT

The different black population groups used to have different taboos about abortion; most were totally opposed to it and would punish a woman who obtained an abortion (Schapera 1940). One proverb states: 'To force out the womb is grievous, the knot of the cradle skin is a flower' (i.e. women derive great happiness from carrying a baby slung on the back). The taboo surrounding the ending of the life of a foetus in utero, however, does not extend to the newborn infant. Infanticide and euthanasia have been practised historically in many situations in Africa. Livingstone (1857) described the case of a mother with an albino child who was excommunicated from her village in Botswana; only when she agreed to put the child to death was she allowed to return. Jeffreys (1953) described the widespread practice of killing one or both twins and sometimes the mother as well, because twins were seen as evil omens who had to be eliminated as quickly as possible. Generally it was the midwife's job to do away with the abnormal child and to tell the father that it was stillborn (Schapera 1940).

In many groups, infanticide was permitted for children with congenital abnormalities, as well as those born with a tooth or born feet first (they were smothered in manure immediately after birth and buried in the hut). When a case of infanticide became known to the chief, it was never pursued or prosecuted (Schapera 1937). This aspect of the culture is changing, however, and attitudes towards children with congenital handicaps are moving towards acceptance. If infanticide is still occurring, it is carried out in secret.

Cultural practices also determined rules about the selection of partners. In several groups the preferred arrangement was with a first cousin. In particular, a man was expected to marry his mother's brother's daughter (Krige 1937). This preference for first-cousin marriages is illustrated in two Tswana proverbs: 'Child of my paternal uncle marry me so that the cattle should return to our kraal' and 'Side by side with his cousin a man is always happy' (Schapera 1940). Consanguineous marriages were motivated, to a certain extent, by the traditional payment of *lobola*, or bride-wealth. If this consisted of cattle, as was often the case, then marriages were often arranged within the extended family if at all possible. Consanguineous marriages were also common among the Swazi (Kuper 1963), but the motive in their case involved keeping the political power within the family. In both of these cases, the cultural practice affected the rates of recessive disorders: albinism for example, is found at a much higher rate in these groups than in other groups (like the Tsonga and Zulu) in which marriage between relatives is taboo (Kromberg & Jenkins 1982). Marriage between relatives was believed to be as common as 30 per cent in the Southern Sotho and Tswana in previous decades, but this figure is falling with the breakdown of ethnic customs.

Polygamy and polyandry (mostly sequential) are practised quite widely in South Africa, and men will quite often have a wife in the rural areas and one in the town in which they work for most of the year. Women, on the other hand, might have several 'husbands' one after the other. From the genetic point of view, these practices have both advantages and disadvantages. Polygamy has been said to alter the sex ratio at birth with an increase in female births (Miller 1993). The polyandrist, however, will only very rarely have two children with the same recessive disease, while, on the other hand, she could have a higher background risk for a recessive condition, since she has her children by several different partners, increasing her chance of mating with a carrier for a disorder for which she is also a carrier. We have therefore, in these situations, had to ask the question 'will you have another child by this father?', and if the answer (for example) is 'no, he's deserted me', then the necessity of emphasising a recessive recurrence risk loses much of its importance. These patterns of relationship have become more prevalent with the breakdown of ethnic structures, reduced use of *lobola* or bride-wealth, the migration to the cities, urbanisation, the increasing opportunities open to women, and the increasing number of female heads of households in urban areas (about 60 per cent in Soweto, Johannesburg).

MYTHS AND SUPERSTITIONS

One of the strongest superstitions we have encountered with regard to genetic disorders is the prevalent view that albinos do not die (Kromberg 1992). This belief is widespread, and we are questioned about it by mothers with newborn affected babies, by albinos themselves, by district health workers, midwives, teachers and many others. Belief in the myth affects the interaction between albinos and the general population at many levels. Mothers tend to reject the child psychologically at birth and only develop normal bonding behaviour when the child is about nine months of age (Kromberg et al. 1987). Later, many show extreme reactions, either spoiling the affected child, some of whom then develop a passive dependent personality, or else rejecting the child, who then may develop an angry, defensive, aggressive approach to life. A discussion of this death myth is an essential part of the genetic counselling offered in this situation (Kromberg & Jenkins 1984). The myth may also cause stigmatisation, which may genuinely be based on community attitudes or may occasionally be self-imposed by the albino himself, who wrongly perceives it in the community.

Another set of relevant myths, about which we still have much to learn, surrounds the mother's beliefs about embryology. It has been reported that, because menstruation stops at the beginning of pregnancy, mothers used to believe that the infant received all its blood from the mother (Schapera 1940). This belief, if it is still prevalent, may make it difficult to explain concepts such as recessive inheritance.

Also, the black woman is still generally blamed by the community for the congenital defect in the child (as is still probably partly true in the white population). On the other hand, in many ethnic groups, a son can only inherit property from his father and a daughter from her mother (Schapera & Goodwin 1937), and so the concept of inheriting genes from both parents may be difficult to comprehend.

LANGUAGE ISSUES

One of the most difficult barriers to effective genetic counselling is associated with language. The local Bantu languages have no words for 'genes' or 'chromosomes', as is probably the case in the languages of most developing countries. The concept of a gene, as a particle of inheritance, frequently has to be explained when genetic counselling is offered, and the English word is often taught to the client. Although we occasionally show illiterate women with Down's syndrome infants a picture of the chromosomes, we do not yet know what they under-stand by them. Post-counselling follow-up is essential in these cases, and although we are providing it only occasionally at present (because of the transport problems for the patient), we are hoping it will soon become more routine.

Other language issues surround words or names, such as Down's syndrome, which are easily misunderstood. This syndrome has, apparently, never been recognised as an entity in any local group, and there is no name for it in the vernacular (Kromberg et al. 1991). The traditional healers, however, give different medicines to affected children compared with those they give to other children who are brought to them (Zwane et al. 1992). We presume this is due to the mothers' reports to the healer, which probably include statements about multiple infections and poor muscle tone in the affected child.

Pedigree-taking can also be complicated in the local situation, since, for example, the client's mother's older sister may be called the client's older mother, instead of aunt, or her younger sister may be called a younger mother. Also, where consanguineous marriages are taboo, the correct information regarding consanguinity may not be forthcoming. On the other hand, strong extended family ties are beneficial in that there is usually available some relative to care for a handicapped child when the parents are not available.

ATTITUDES TO REPRODUCTION AND THE VALUE OF CHILDREN

Several attitudes handed down in the culture of various groups affect reproduction, child-bearing and child-rearing practices. A woman's status is frequently established by the number of children she has, especially in the rural areas. Family planning and contraception are therefore unpopular primary health care services (although women may be more in favour than men, who may expect their wives to have a child at regular intervals); in one of our recent studies on the Tsonga, 80 per cent of the women had never used contraception (Kromberg, unpublished material). There is also a high infant mortality rate (IMR) and the concomitant need for replacement children; if a woman wants to be sure of some children surviving to look after her in her old age, she will continue having children up to menopause. The value of children may be affected by the poverty level of the mother, and several of our women clients have verbalised the need to have a child (consequently refusing prenatal diagnosis when it was strongly indicated) to persuade the father to support both the woman and the child. This is particularly so in cases of advanced maternal age, where the women does not have a permanent stable relationship with a man.

The high IMR, however, also leads to the attributing of relative unimportance to hereditary disease, from three points of view. First, survival of the fittest is occurring, which generally means that the children with congenital and genetic disorders often do not survive and therefore these conditions are not recognised as a sizeable problem (we know, for example, that approximately 1 in 600 infants is born with Down's syndrome (Kromberg et al. 1992), but from our observations there are very few such infants alive in the community). Second, the attention of primary health care providers is centred on the major childhood diseases, such as gastro-enteritis, respiratory infections and malnutrition, so that congenital and hereditary disorders come low on most lists of causes of death in childhood. Third, the high IMR must affect attitudes, and in Gazankulu (Eastern Transvaal), where the IMR is about 120 per 1000 births, women probably adopt a more 'matter of fact' attitude towards death and risks, and when genetic counsellors give a 1 in 4 risk for a serious genetic condition, this is not perceived as being significant in the context of the background risk.

Changing cultural expectations of women themselves, however, are slowly starting to affect the number of births per woman, and this is now halved in urban areas (average 2.8 per woman), where infant mortality rates are much lower (22 per 1000 in Johannesburg) compared with the number of births per rural women (5.7 in deep rural areas, 4.2 in semi-rural areas) (Cooper et al. 1993).

USE OF WESTERN AND TRADITIONAL HEALERS

The use of both Western medical and traditional healer services is widespread; in a recent study we found that 48 per cent of mothers of Down's syndrome children claimed to use both services (Zwane et al. 1992). This practice means that the genetic facts can be totally discarded if the traditional healer gives a different opinion. One case illustrates this sometimes disastrous turn of events. A couple with two children affected by xeroderma pigmentosum presented for genetic counselling. They were told of the 1 in 4 recurrence risk and the availability of prenatal diagnosis. They asked for the procedure and underwent it, at great expense to the state (foetal samples went to the USA). The couple were told the foetus was affected, but they consulted the traditional healer who disagreed. The pregnancy continued and they had an affected child; some years later they had another affected child, but prenatal diagnosis was not requested in subsequent pregnancies. The power of the traditional healer is still not fully understood and requires further study. Investigation is not easy as the power of the traditional healer's treatment is often related to the secrecy surrounding the whole consultation, and it is therefore difficult to obtain reliable information from either the traditional healer or his or her patients.

CONCLUSION

In this paper we have described and discussed some of our experiences relating to cultural issues and genetics. Many of these issues require more systematic scientific study, but we raise them here in order to make workers in the field aware that there are still many pitfalls in cross-cultural counselling, especially in the practice of genetic counselling, in Southern Africa.

ACKNOWLEDGEMENTS

We wish to thank the South African Medical Research Council, the SA Institute for Medical Research, the Richard Ward and the Iris Ellen Hodges Funds of the University of the Witwatersrand, who have provided support for our work over the years. We also acknowledge the assistance of Janet Robbie for patiently and expertly typing the manuscript.

REFERENCES

Cooper, C., Hamilton, R., Mashabela, H., MacKay, S., Sideropoulos, E., Gordon Brown, C., Murphy, S., & Markham, C. 1993 *Race Relations Survey 1992/93.* SA Institute for Race Relations: Johannesburg.

Hammond-Tooke, D. 1989 *Rituals and Medicine.* A.D. Donker: Johannesburg.

Jeffreys, M.D.W. 1953 Twin births among Africans. *South African Journal of Science* 50: 89–93.

Jenkins, T. 1990 Medical genetics in South Africa. *Journal of Medical Genetics* 27: 760–79.

Jenkins, T., Wilton, E., Bernstein, R., Nurse, G.T. 1973 The genetic counselling clinic at a children's hospital. *South African Medical Journal* 47: 1834–8.

Kluckhohn, C. 1965 *Mirror for Man.* Fawcett World Library: New York.

Krige, E.J. 1937 Individual development, in Schapera, I. (ed.) *The Bantu-Speaking Tribes of South Africa.* Maskew Miller: Cape Town, pp. 95–118.

Kromberg, J.G.R. 1992 Albinism in the South African negro. IV: Attitudes and the death myth, in Evers-Kiebooms, G., Fryns, J.P., Cassiman, J.J., & van den Berghe, H., (eds) *Psychosocial Aspects of Genetics.* Birth Defects Original Article Series 28. Wiley: New York, pp. 159–66.

Kromberg, J.G.R., & Berkowitz, D. 1986 Managing genetic disorders. *SA Family Practice* 7: 279–85.

Kromberg, J.G.R., Christianson, A.L., Duthie-Nurse, G., Zwane, E., & Jenkins, T. 1992 Down syndrome in the black population. *South African Medical Journal* 81: 337.

Kromberg, J.G.R., & Jenkins, T. 1982 Prevalence of albinism in the South African negro. *South African Medical Journal* 61: 383–6.

Kromberg, J.G.R., & Jenkins, T. 1984. Albinism in the South African negro. III: Genetic counselling issues. *Journal of Biosocial Science* 16: 99–108.

Kromberg, J.G.R., Zwane, E.M., Duthie-Nurse, G., & Jenkins, T. 1991 Down syndrome in a 3rd world community. *American Journal of Human Genetics* 49 (supplement): 268.

Kromberg, J.G.R., Zwane, E.M., & Jenkins, T. 1987 The response of black mothers to the birth of an albino infant. *American Journal of Diseases of Children* 141: 911–16.

Kuper, H. 1963 *The Swazi: A South African kingdom.* Holt, Rinehart and Winston: New York.

Livingstone, D. 1857 *Missionary Travels and Researches in South Africa.* John Murray: London.

Manganyi, N.C. 1973 *Being Black in the World*. Ravan Press: Johannesburg.
Manganyi, N.C., Kromberg, J.G.R., & Jenkins, T. 1974 Studies on albinism in
   the South African negro. I: Intellectual maturity and body image differen-
   tiation. *Journal of Biosocial Science* 6: 107–12.
Miller, S.K. 1993 Many wives means many daughters. *New Scientist* 1876: 5.
Mofokeng, Z. 1991 Attitudes of black women toward mental retardation and
   termination of pregnancy for fetal abnormality. B.Sc. (Hons) dissertation,
   Department of Human Genetics, University of the Witwatersrand, Johan-
   nesburg.
Read, M. 1966 *Culture, Health and Disease*. Tavistock: London.
Schapera, I. 1937 Law and justice, in Schapera, I. (ed.) *The Bantu-Speaking
   Tribes of South Africa*. Maskew Miller: Cape Town, pp. 197–219.
Schapera, I. 1940 *Married Life in an African Tribe*. Faber and Faber: London.
Schapera, I., & Goodwin, A.J.H. 1937 Work and wealth, in Schapera, I., (ed.)
   The Bantu-Speaking Tribes of South Africa. Maskew Miller: Cape Town, pp.
   131–71.
Zwane, E.M., Duthie-Nurse, G., & Kromberg, J.G.R. 1992 Down syndrome
   in the black population: diagnosis, Western and traditional treatment pro-
   blems. Abstracts. International Association for the Scientific Study of Mental Deficiency, 9th World Congress, 3–9 August 1992, Australia.

11 The Role of Unconscious Fantasy in the Giving and Receiving of Genetic Counselling

Gulsan Karbani, Susie Godsil and Robert Mueller

This chapter is written from the perspective of those engaged in genetic counselling in an area of Britain with great cultural and ethnic diversity. Our aim is to focus attention on aspects of the genetic counselling process which can easily be overlooked as we strive to improve the ways in which we deliver complex genetic information. We are interested in the way in which patients' fantasies about a genetic disease, about the information given and about the course of action taken, may play a part in this process. We are also interested in how clinicians' fantasies about white and black cultures may influence the way they interact with patients.

Our thoughts derive from psychoanalytic theories of the unconscious, that human beings operate only partially on a conscious level and that much of our emotional life is unconscious to us, particularly fantasies relating to sex and reproduction, aggression and death. Sigmund Freud's description of how children often deal with sex education and the facts of life might be useful to bear in mind when we think of patients faced with factual genetic information: 'They behave like primitive races who have had Christianity thrust upon them, and who continue to worship their old idols in secret' (Freud 1937).

It is interesting that Freud, of Jewish origin himself, was using this image of (white, Western) missionaries trying to convert (reluctant) primitive peoples to Christianity, in order to describe universal aspects of mental functioning. One can see in his choice of analogy, however, how readily people of other cultures may come to represent the primitive aspects of us all. From the perspective of white, middle-class professionals in Britain, the 'other' cultures to be labelled as primitive will be those associated with the peoples of Britain's former empire; 'black' cultures may be seen as the products of a mass of submerged, primitive impulses. Thus, universal aspects of mental functioning may be seen (falsely) as being split between cultures, between black and white, instead of (more appropriately) being dealt with as part of each and every one of us (Da Conceicao Dias & De Lyra Chebabi 1987).

This association between the two concepts 'black' and 'primitive' may have repercussions in clinical work. Black patients may be immediately assumed to have various superstitious ideas about genetic information. White, particularly middle-class, patients may be assumed to have a more rational understanding of such matters. Educated patients, white or black, may feel ashamed of some of the fantasy theories they have in their minds about a disease, and so hide them from the counsellor. This may particularly affect educated black patients, who may feel an extra shame about what they experience as primitive thoughts. These factors may affect what sort of service provision is offered and to which patients, and how these matters are explored with particular patients. We shall elaborate these hypotheses by providing case examples followed by a discussion of the issues raised.

EXAMPLE 1

Family A was referred by the consultant community paediatrician to the Department of Clinical Genetics as there was concern regarding two boys with severe learning difficulties and very challenging behaviour. The ten-year-old boy had moderately severe learning difficulties with less severe behaviour problems. A range of medical investigations (including chromosome analysis and testing for fragile X syndrome) were all normal. He was the elder of the two boys; he was born in Saudi Arabia and was apparently kept in the special care baby unit there for two weeks because he was very small and sick in the neonatal period.

The five-year-old child, who was born in Pakistan, apparently walked some time after 12 months of age, and talked shortly after two years. When his father received a scholarship to study overseas, the child was very upset and stopped talking soon after the father left and has never spoken again. Investigations – including cranial CT scan, amino acid chromatography, mucopolysaccharide screen,

thyroid function tests, and serum calcium, bioprofile, white cell lyso-
somal enzyme and chromosome analysis – have all been normal, and
there was no evidence of fragile X.

The referral letter had described the family as being very anxious,
and in such instances normal departmental practice was to arrange a
pre-clinic home visit by a genetic counsellor. The paediatrician was
concerned that the father, who was a well-respected leader in the
mosque, would not welcome the counsellor because she was a
woman. The counsellor gave the matter some thought, encountered
similar worries in her own mind, but felt that a concern was being
created derived from an unhelpful stereotype in the paediatrician's
mind. There seemed little basis for these assumptions in reality, and
so the counsellor went ahead with the visit as planned.

The family were very welcoming and appreciated the concern shown
by the counsellor. The father was an educated man with a broad grasp
of genetics. He demonstrated a readiness to talk about his feelings and
fantasies about the origins of his sons' conditions. Whilst he felt that
his elder son's difficulties were probably genetic, he had particularly
strong fantasies about his five-year-old. He felt he had been a great
object of envy in his community when a second son was born and
when he won the scholarship to study abroad. He worried that when
he left Pakistan, he had left his family unprotected and exposed to the
envious curses of those around. Along with this worry, he also showed
some psychological insight into the impact his absence may have had
on this son, who had been extremely attached to him; he talked of his
terrible feeling of guilt at having left him in this state for five years,
particularly since his wife had kept him in ignorance of the boy's
muteness, not wanting to worry him.

At the subsequent appointment in the clinic, the consultant
explained that there were indeed genetic implications regarding the
elder boy's behaviour and appearance, but he did not believe this to be
the case with the younger child. The father agreed with the consultant's opinion and referral to child psychology was arranged.

DISCUSSION

As we have stated above, pre-clinic visits are automatically offered in
our service when the referral letter makes specific reference to the
patient's or family's anxiety or when there is recent emotional trauma,

such as a death or a new diagnosis. There is the ordinary intention to offer some sort of support to the family, particularly when it may be some time before a clinic appointment, but the offer is also based on the ideas implicit in our earlier introduction. These anxious situations tend to stir up all sorts of worrying thoughts and fantasies, and the opportunity to explore these with a counsellor can leave the patient in a better position to make use of the appointment with the doctor. In our example, it is clear how easy it might have been to deprive family A of this service. We could speculate further about this. For instance, what might appear as a genuine mark of respect for another culture's assumed attitudes to women could also be a projection of the paediatrician's own disavowed sexism and undervaluing of the 'woman's work' of counselling as opposed to the 'man's work' of doctoring. The ancient hierarchy of doctors and nurses dies hard!

We also see something of the assumption that black patients will automatically have superstitious, closed ideas; this was raised, in this case, by the fact that the father was an imam. The counsellor concerned is herself a Muslim and was thus less vulnerable to making these assumptions. Although she felt some worry within herself as to how she would be received, this is not an unusual state of mind before the first visit to a family; she could stay in touch with the reality that whether this family could use some help or not would only be discovered by proceeding as normal.

As can be seen from our short account of the visit, the family – and the father in particular – were hungry for a space to talk. What is interesting and important to notice is how this example so clearly illustrates the way in which reality-based thinking and primitive fantasies can coexist. Although this is not a paper on technique, we would stress the role of the counsellor here in encouraging the father to express all his 'theories' about his sons and to elaborate on them, rather than closing them down too quickly or rationalising them away.

The father seemed to feel relieved that there was someone who might understand his feelings of having left his family exposed to the 'evil eye'. He had felt unable to talk of such things to the white GP or paediatrician but felt that the Muslim counsellor would understand. Clearly, at this point, he is projecting his rational self into white culture as though white culture had no concept of spite or envy.

When given the freedom to express these paranoid fantasies, however, the father seemed to be able to move of his own accord to a position with more psychological insight, which involved him connecting with guilty feelings about the impact of his absence on his child.

Although this was very painful, it was more reality-based and it allowed him to reclaim his own capacity for rational thought. Prior to this visit, the father was confused regarding the problems with both his sons. It seems that the pre-clinic visit enabled him to disentangle his primitive fantasies about the meaning of his sons' difficulties from the genetic and psychological aspects of the children's behaviour, thus allowing appropriate treatment and referral to take place.

EXAMPLE 2

Mrs B. is a white, middle-class woman in her 30s, working as a senior librarian. After years of unsuccessful treatment for infertility, she and her husband finally conceived by ordinary methods. It was a twin pregnancy, and at the 18-week scan, both babies were found to have been dead for several days. Mrs B. was without her husband at the clinic and was devastated by this terrible information. The genetic counsellor who was present sat with her until her husband arrived and sustained the full force of this woman's rage, distress and disappointment over what had happened – not only about the current situation but also all the dashed hopes endured over the years with in vitro fertilisation (IVF) treatments.

After the delivery of the twins, Mrs B. was offered and attended a series of counselling sessions with the same counsellor. At first she seemed to want to relive the trauma, going over the details of the scan, the labour, the delivery of the dead babies, and the experience of talking to the consultant, the radiographer, the counsellor and the midwife. She went through the experience again and again, bombarding the counsellor with her grief and her rage. She felt unable to express her feelings with her husband, family, friends or colleagues and was terribly ashamed of what she was saying.

It became clear eventually that she still feared to raise certain topics with the counsellor. With encouragement, she started to speak of the desperate urge she had to dig up the babies when she visited their graves. She wanted to dig them up, hold them in her arms and tell them how much she missed them. She had been keeping such fantasies to herself for fear that people would think she was mad. As she put these thoughts into words, she became connected with her guilty feelings at not having held the twins when they were born, and she remembered how angry she had felt with them for having rejected her by dying.

Soon after this, she had a dream in which a close friend, who had recently died, was sitting in an armchair by a fire, holding her twin babies in his arms. She experienced the dream as reassuring, as though her babies were well looked after and in safe hands. Although she had vowed never to think of another pregnancy after the death of the twins, she gradually renewed contact with the IVF unit and started to think about another baby. Twenty-one months later, the counsellor received a letter telling of the safe arrival of a baby boy conceived without IVF.

DISCUSSION

We wished to include a clinical example from work with an educated, white patient to emphasise that primitive fantasies may arise in any patient, and that we see no connection between this and race, class or capacity for rational thought. As we state in our introduction, aspects of mental functioning can be seen as being split between cultures – or being attributed to different cultures – instead of being recognised and dealt with as part of each one of us.

In our first example, this could have led to an assumption that the patient would have no capacity for rational thought. In the case of Mrs B. and others like her, our experience is that such patients are often presumed not to need counselling help, as if their 'brains' will see them through. This is complicated by the patient's own view that such thoughts are shameful. Shame played a large part in Mrs B.'s experience of how the loss of her babies affected her. The counsellor's capacity to sense this and attempt to make contact with these hidden wishes allowed Mrs B. to put these thoughts into words, and eventually to move on to lay the twins to rest in her mind. It was almost as though the airing of the mad wish to dig up her babies made it possible for her to stop endlessly digging them up again and again in the counselling sessions.

CONCLUSION

In this paper we have wanted to demonstrate the usefulness of a genetic counselling approach which takes into account the way primitive fantasies are stirred up in patients, reducing their capacity to think in a reality-based manner. We have wanted to show how this is further complicated in a trans-cultural setting, if we are unaware of the tendency to attribute a primitive mentality to black people and a civilised mentality to white people, ascribing different modes of thought to each 'culture'.

In order to make proper use of genetic information, patients must be able to think in a reality-based manner. Genetic information is never neutral. With its central image of something bad inherited from our ancestors and potentially passed on to our children, clinical genetics can stir up primitive fantasies which are usually unconscious and can deeply affect a patient's capacity to think clearly. This is complicated by our sense that, alongside the shame often felt about genetic illness, there can also be a shame about having such 'crazy' thoughts and feelings about it. Our contention is that we can learn to acknowledge the presence of these old idols in our own and our patients' minds. In this way, we may assist in the proper assimilation of factual information in clinical genetic consultations, and thereby promote an appropriate adjustment to reality.

BIBLIOGRAPHY

Da Conceicao Dias, C.G., & De Lyra Chebabi, W. 1987 Psychoanalysis and the role of black life and culture in Brazil. *International Review of Psychoanalysis* 14: 185–202.
D'Ardenne, P., & Mahtani, A. 1989 *Transcultural Counselling in Action.* Sage: London.
Freud, S. 1937 *Analysis Terminable and Interminable.* Standard Edition of the Complete Works of Sigmund Freud, vol. 23. Hogarth Press and the Institute of Psychoanalysis: London, pp. 216–53.
Jacob, M. 1989 *The Presenting Past: An introduction to practical psychodynamic counselling.* Sage: London.
Stewart, W. 1992 *A–Z Counselling Theory and Practice.* Chapman and Hall: London.
Wag, V., & Marsh, F.M. 1992 Ethical principles and cultural integrity in health care delivery: Asian ethnocultural perspectives in genetic services. *Journal of Genetic Counselling* 1(1): 81–92.

Part III
Inheritance

12 Everyday Ideas of Inheritance and Health in Britain: Implications for Predictive Genetic Testing

Charlie Davison

In this chapter I am presenting an analysis of data collected in three research sites in South Wales in 1988–90. Qualitative interviews were carried out with a randomly selected sample of adults (N=180). These individuals were drawn from the electoral registers of three quite different electoral wards. Plasnewydd (an inner-city, socially mixed part of central Cardiff), Porth (an industrial and service town in the Rhondda Borough) and Llangammarch Wells (a village and rural area in the Borough of Brecknock). The sample was constructed to ensure a spread of ages and social classes, as defined by the Registrar-General's occupational classification. Both men and women were interviewed. (See Table 12.1.)

Table 12.1   South Wales cultural aspects of ischaemic heart disease project: semi-structured interview respondents (Jun. 1988 – Dec. 1989).

R-G occ. class	Males				Females				All
	Pl.	Po.	Ll.	All	Pl.	Po.	Ll.	All	
I + II	9	6	9	24	9	8	9	26	50
III n	8	2	4	14	8	9	6	23	37
III m	9	8	9	26	7	7	8	22	48
IV + V	9	5	9	23	4	9	9	22	45
All	35	21	31	87	28	33	32	93	180

The interviews with the sample formed the 'back-bone' activity of six months fieldwork in each area. During that time, many opportunistic conversations and observations contributed to a more general process of less structured ethnographic data collection. Both the more formal interviews and the less structured work focused on explanations and representations of chronic disease, especially ischaemic (coronary) heart disease (IHD) and the popular culture of prevention. Heredity was identified as a major theme during fieldwork. While the semi-structured interview schedule included no questions about heredity, many informants raised the issue spontaneously. This was mainly in response to the question 'why do you think it is that some people get ill and others don't?', or in discussing some aspect of their own health experience. When the issue arose, it was sometimes mentioned in the same breath as an assessment of whether an individual had control or influence over the hereditary aspects of health. Table 12.2 presents information about where and by whom the topic of heredity was raised, indicating the very wide distribution and recognition of heredity as an important health issue.

Table 12.2 Social class of informants spontaneously mentioning heredity as an important factor in determining who becomes ill and who does not.

	Men			Women		
	Can control	Cannot control	Mention	Can control	Cannot control	Mention
I + II	3	3	13	5	1	12
III n	3	0	9	1	3	14
III m	0	4	12	1	0	11
IV + V	2	0	8	0	0	10

115 out of 180 overall: 63.9%

Gender
57 out of 87 men: 65.5%
58 out of 93 women: 62.4%

Registrar-General's occupational class
67 out of 87 non-manual (I, II, III n): 77.0%
48 out of 93 manual (III m, IV, V): 51.7%

As this table shows, there is a distribution across gender and social class groups. Furthermore, some informants spontaneously gave views on whether inherited attributes were or were not open to control or modification by the person inheriting them.

It was the aim of the project to gather data on the popular culture of the transmission of traits across generations. Analysis of this material has been carried out in the context of a more general interest in knowledge about possible future health events and ways of integrating such knowledge into daily life. It is these data that are to be discussed in this chapter, although space does not permit a comprehensive treatment of the issues raised.

While heredity was mentioned and discussed by many of the informants in the study, it should be noted that this does not necessarily mean that their opinions, experiences or beliefs, or the importance of their thoughts about heredity were uniform. About kinship in general, Strathern has pointed out that:

> Anyone embarking on a study of English kinship practices would certainly feel bound to specify the class background of their study, and would expect to be dealing with distinctive class features. (Strathern 1992: 23)

The same is no doubt true of the more restricted field of inherited aspects of health. I believe, however, that it is possible to discuss health inheritance, and the wider kinship landscape in which it is set, at a level of abstraction that allows us to talk, in general terms, about the existence of 'a system'. For the purposes of this chapter, I want to use our study of three communities in South Wales to generalise outward and think about a British (or at least a 'white British') kinship.

PREDICTIVE GENETIC TESTING

The phrase 'predictive genetic testing' refers to the examination of a sample of genetic material with the aim of producing information about the health-related future of the person from whom it was taken. Such tests can be carried out on genetic material collected from individuals as foetuses or at any age after birth. At present, predictive genetic testing has several routine applications in antenatal care, such as the use of chorionic villus sampling in the early identification of foetuses with chromosomal abnormalities. The use of predictive genetic testing in people already born has, until recently, been restricted to a handful of relatively rare genetic disorders, such as Huntington's disease or muscular dystrophy. This paper, however, is principally concerned with the extension of predictive testing into a wide range of very common illnesses and conditions which is being facilitated by current rapid developments in human genetics (see, for example, Wilkie 1993).

In the vast majority of settings in the modern world, a normal fact of social existence is that most people have a notion of who they are in terms of where they came from (in the technical sense that some aspects of their social identity are derived from their membership of a descent group). This rudimentary aspect of kinship systems does not suggest, of course, that socially defined descent is the same as biologically defined descent; indeed, the two often vary (for a classic anthropological debate on this, see Gellner 1987: 154–82, and Needham 1960). This variation exists not only on the theoretical level, but also on the practical, in that it is estimated that quite large numbers of British people (up to 10 per cent) are not the biological offspring of the couple they have always thought of as their mother and father (Macintyre & Sooman 1991).

One aspect of reckoning shared by contemporary white British systems of analysing descent, however, is the idea that any individual belongs to a family with different 'sides'. This notion arises from the fact that, in common with many other systems with European origins or influences, British kin reckoning tends to be bilateral, which is to say that an individual belongs to both their mother's and their father's family groups.

The symbolic importance of 'sides' is underlined by their use in the physical organisation of the traditional British wedding service, the ceremony bestowing social and cultural legitimacy on a reproductive unit. Here the church's central aisle is used ritually to separate the family and supporters of the bride from those of the groom. This spatial layout thus maps the two sets of original nature/nurture influences that any future offspring of the nuptial couple will refer to as 'my mother's side' and 'my father's side'.

The interaction between the kinship system and the application of clinical genetics hinges on the fact that British people expect to inherit a whole range of traits, characteristics and attributes from their forebears, who are organised conceptually into 'sides'. (See Table 12.3).

In terms of the charting of the inheritance of ailments, weaknesses, susceptibilities and the physical and personal attributes that may be associated with them, British 'sides' systems have much in common with the descent models of clinical genetics. There is a common expectation that children will resemble parents and grandparents. For such 'vertical' transmission through generations, British people use the term 'taking after'. Such similarities are also discernible between related people of the same generation, however, and for this 'horizontal' axis of the system, the terms 'spit' or 'image' are more commonly used.

Table 12.3 What do people inherit?

1. Noticeable physical attributes:
   eye colour, hair colour/type, face shape, 'build', weakness of parts of body (lungs, chest, eyes, heart), tendencies to specific ailments (varicose veins, asthma), specific ailments (Parkinson's, 'blood conditions', high blood pressure).

2. 'Constitution':
   general resistance to ailments ('strong' or 'hard'), general tendency to sickness ('weak'), longevity (or not), metabolism (tendency to put on weight, skinny glutton).

3. Personality:
   disposition ('sunny', 'negative', 'laid back'), specific attributes (being a 'worrier', talkative, talented, fear of heights, sweet tooth), behavioural styles (hair flicking, gait, facial expression).

Because the 'sides' system implies a dendritic structure (my parents also each came from two sides, so did theirs, etc.), the charting of traits and attributes even within a relatively limited generational depth takes on an aspect of speculation and even lottery (Davison et al. 1989). Further complication is added by the tradition that 'inheritance' can either be essentially physical or essentially social in character. In the everyday discourse that surrounds these issues, physical and social inheritance form two quite distinct conceptual sets, relating to the contrast between nature and nurture:

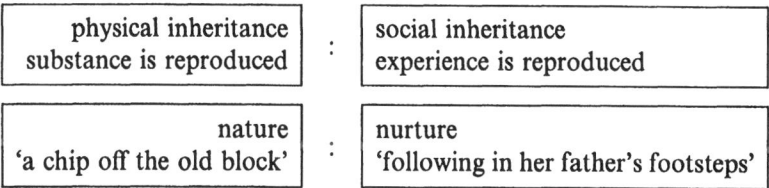

physical inheritance	social inheritance
substance is reproduced	experience is reproduced

nature	nurture
'a chip off the old block'	'following in her father's footsteps'

This conceptual model (and its attendant idiom) of the transmission of physical and social characteristics through generations is readily extended into the world of personal things or possessions. Thus, British people can speak equally readily and equally accurately of the 'inheritance' of both ingrowing toe nails and diamond rings.

Because of the superficial similarity between scientific genetics and the 'sides' aspect of British bilateral kinship reckoning, it is tempting to believe that the findings of predictive genetic tests will be relatively easily assimilated into family life. Such an expectation is illustrated by the fact that genetic counsellors tend to see the communication of accurate probability and odds calculations (rather than conceptual issues concerning inheritance) as their main professional role (Harper 1988, Parsons & Atkinson 1992).

In terms of the interaction of predictive genetic testing with popular cultures, two issues are of particular importance. First, British popular culture expects to find character traits, such as nervousness or wit, following the same sort of route as hair or eye colour. Second, popular systems are much more ready to link traits and attributes into significant groups, giving primacy to a relatively 'holistic' appreciation of self and inheritance (Davison et al. 1994). The importance of this divergence in terms of the impact of predictive genetic testing is best illustrated by an example.

The daughter of a family, as she moves into her late 30s, is recognised by relations, friends and neighbours alike as developing a build (thick arms, strong shoulders) and a gait (slightly rolling) that make her the 'image' of her grandmother at that age. She has also acquired the habit of flicking the hair from her eyes that makes her just like her mother. The fact that she has started to (absent-mindedly) miss sections of conversations and then interject inappropriate opinions is so like her mother that it is the cause of much mirth amongst family and friends. Because of these inheritances, it is widely assumed that the woman in question will very probably suffer with arthritis in her late 50s and 60s because so many of the women ('on both sides, actually') in the family have – and particularly because the mother and the grandmother (so strongly 'taken after' in other ways) both had a lot of trouble with their joints in later years. Closer family members have quite serious concerns about the severe depression suffered by the woman's mother when she was in her mid-30s and again at menopause.

In this example, a range of different physiological and personal traits are charted through three generations. For the protagonists and observers, they make a set, and it 'stands to reason' that the inherited body shape, the absent-mindedness and the slightly humorous conversational quirk should go together. This easy relationship with linkage leads naturally to the conclusion that arthritis is a distinct possibility, and that depression may never be far away. An extreme form of such processes (termed 'pre-selection') has been charted in families with Huntington's disease (Kessler 1988).

The example can be hypothetically extended into a time in which predictive genetic testing is available and the woman, as a youngster, is provided with a personal genetic profile. As this service is part of the health system, information given is concerned with vulnerability and susceptibility, and the emphasis is placed on the benefits of early prophylactic action. The principal finding, in this case, is that a marked susceptibility to arthritis is identified, and this information is passed on with the recommendation that great care should be taken with respect to choice of sports activities, heavy lifting and repetitious arm and hand movement (in a job, for example).

When this information is placed, as an overlay, onto the popular culture of inheritance, the linkage process is played out in reverse. The intimation that arthritis is 'on the cards' is now a starting point and may stand as official scientific evidence that the woman 'takes after' her mother and grandmother. Because family inheritance lore tends to link joint trouble with depression, a strong potential exists for the advanced 'labelling' of the testee as someone to be treated in a particularly kind (or harsh) manner. While predictive genetic testing in Huntington's disease had the potential to prevent damage to a person through inaccurate 'pre-selection' (Kessler 1988, Kessler & Bloch 1989), its application more generally may have the opposite effect. Because of the relative complexity and sophistication of the 'sides' system, medically produced knowledge (which will inevitably consist of a series of 'bits' of information) will be easily assimilated. But through the action of popular 'linkage' concepts, giving additional associations to the factual content itself, the results of predictive genetic testing will often mean more to family members than those who produced and communicated the information intended.

REFERENCES

Davison, C., Frankel, S., & Davey Smith, G. 1989 Inheriting heart trouble: the relevance of common-sense ideas to preventive measures. *Health Education Research – Theory and Practice* 4: 329–40.
Davison, C., Macintyre, S., & Davey Smith, G. 1994 The potential social impact of predictive genetic testing for susceptibility to common chronic diseases: a review and research agenda. *Sociology of Health and Illness* 16 (3): 340–71.

Gellner, E. 1987 *The Concept of Kinship*. Blackwell: Oxford.

Harper, P.S. 1988 *Practical Genetic Counselling*, 3rd edition. Wright: London.

Kessler, S. 1988 Preselection: a family coping strategy in Huntington's disease. *American Journal of Medical Genetics* 31: 617–21.

Kessler, S., & Bloch, M. 1989 Social system responses to Huntington's disease. *Family Process* 28: 59–68.

Macintyre, S., & Sooman, A. 1991 Non-paternity and prenatal genetic screening. *Lancet* 338: 869–71.

Needham, R. 1960 Descent systems and ideal language. *Philosophy of Science* 27: 96–101.

Parsons, E.P., & Atkinson, P. 1992 Lay constructions of genetic risk. *Sociology of Health and Illness* 14: 437–55.

Strathern, M. 1992 *After Nature: English kinship in the late twentieth century.* Cambridge University Press: Cambridge.

Wilkie, T. 1993 *Perilous Knowledge: The Human Genome Project and its implications.* Faber and Faber: London.

13 It Runs in the Family: Lay Knowledge about Inheritance

Martin Richards

Genetics concerns families and kinship. It is the study of the ways in which inherited characteristics are passed from parents to children through the generations. As the new genetics (genetics based on recombinant DNA technology) develops, the ability to characterise the genes that individuals may carry increases dramatically. Almost daily there are announcements of the identification of genes related to specific disorders. This allows new possibilities of genetic testing. So, for example, for many genetic disorders, a foetus may be tested and aborted if it is found to carry the disorder that runs in the family. Or in the case of diseases that have an onset in adulthood, individuals can be tested before symptoms develop and so discover whether or not they are likely to get the disorder. There are many questions about such testing and how it should be deployed (see Marteau & Richards 1996 for a more general discussion).

Our approach to these issues has been to investigate the knowledge and beliefs that family members may hold about genetic disorders that may run in their families, and more generally about inheritance, and to describe how information about such matters is communicated among family members. We have suggested that lay beliefs about inheritance, which may be incompatible with a Mendelian view, are a major determinant of reproductive choices, uptake of genetics services and other actions of members of families which carry a genetic disorder (Green et al. 1993, Richards 1993, Richards & Green 1993, Richards et al. 1995).

Our work does not involve cross-cultural comparison (at least in its current phase) and so might not seem very directly related to the theme of this volume. I want to argue to the contrary, however, as it is my hypothesis that lay knowledge about inheritance is closely related to ideas about kinship and the social relatedness of family members. If this proposition is correct, lay knowledge about inheritance would be expected to vary between cultural groups with their different concepts about kinship, marriage and family relationships more generally.

Another point needs to be made about cultural diversity. In genetics, as in many other spheres, the issue of cultural diversity is usually addressed by looking at minority groups whose culture differs from that of the mainstream. Thus we find studies of minorities who have particularly high frequencies of certain genetic disorders (such as sickle cell disorders in Afro-Caribbean communities or Tay-Sachs disease in Jewish communities) or have cultural patterns, such as cousin marriage, that may be held by some to constitute a genetic problem. Such approaches take for granted, or regard as normal or usual, the knowledge, beliefs and practices of the mainstream culture. But, of course, the mainstream culture is as much a part of ethnic diversity as the beliefs and practices of any other group. It is part of our research programme to examine knowledge about inheritance and its possible association with concepts of kinship within our own culture. We would expect the descriptions we can provide to be culturally specific, and similar research is needed for other cultures, including those of ethnic minority communities in the UK.

My aim in this chapter is to sketch out some of what we know about certain genetic disorders from a family point of view (for a further discussion, see Richards 1996) and then discuss the issue of lay knowledge of inheritance. As yet, research is limited, and much of what I say is necessarily speculative, more in the way of a research agenda than a research review. However, one clear conclusion from what has been done so far is that such issues can be very different for different genetic conditions depending on such matters as the nature of the disorder, the age at which symptoms develop, treatments that may (or may not) be available, the mode of inheritance, the availability of genetic tests, and the penetrance of the gene (the likelihood of individuals who have inherited the faulty gene developing the condition). This variation makes it very difficult to make points relevant to genetic disorders in general. I shall discuss two specific disorders to illustrate certain points.

HUNTINGTON'S DISEASE

Huntington's disease is a degenerative disease of the central nervous system which usually first develops in middle age. Its symptoms include involuntary movements (chorea – hence the earlier name Huntington's chorea), loss of motor control and dementia. The disease is progressive and invariably fatal, usually several years after the first appearance of marked symptoms. There is no cure; at best some not very effective palliative treatments are available for some of the symptoms. It is thought to occur with a frequency of about one per 5000 births in Britain (see Weatherall 1991). The disease is inherited dominantly so that those with one copy of the faulty gene (and one normal gene) will develop the disorder. Probably all those with the faulty gene will develop the disease provided they live long enough for it to develop. The disease is exclusively genetic in the sense that it is thought never to occur in those who do not have the faulty gene.

Since the early 1980s, the location of the faulty gene on chromosome 4 has been known and genetic markers have been available for linkage testing. This means that, provided DNA samples are available from a number of family members with and without the disorder (and so the faulty gene), it is possible to tell whether individuals in the family have a high or low probability of carrying the gene by a comparison of their DNA with that of other family members. This test can be done at any stage of life and so can be used as a foetal diagnosis or later in development if individuals wish to know whether or not they are going to develop the disease. The story of the development of linkage testing and its deployment is well described by Nancy Wexler (1992), who comes from a family which carries Huntington's disease and was involved in the research which led to the location of the gene. A number of those who have considered being tested have reported their experiences (see Wright, Madigan & Anon in Marteau & Richards 1996).

As the faulty gene is dominantly inherited, children of a parent who has the disorder have a 50 per cent chance of having inherited the gene and so of getting the disease. Such children, when adult, can (in many cases) use linkage testing and the abortion of an affected foetus to ensure that they only have unaffected children. However, if an affected foetus is found, it implies that the parent must have the faulty gene. Testing procedures have been developed which avoid revealing the parents' own genetic situation, but at the cost of a 50 per cent risk of aborting a normal foetus. In essence, this is done by determining whether the foetus has inherited the relevant chromosome from the normal or affected grandparent.

In 1993 the gene itself was identified (Huntington's Disease Collaborative Research Group 1993). This means that direct testing is now possible – that is, an individual can be tested for the presence of the faulty gene without the need to collect DNA samples from other family members (saving the complication that such tests may be inconclusive in a very few cases).

How do families cope with this devastating disease and its inheritance? There are a number of descriptions of the ways in which the disease has direct and indirect effects on many members of a family. One such vivid description (Klein 1982) appears in the biography of Woody Guthrie, but this comes from the era before any genetic testing was available. In the phase when testing was a possibility but had yet to arrive, it was widely predicted on the basis of surveys of family members and geneticists that about 75 per cent of those who were children of a sufferer, and so had a 50 per cent risk of inheriting the disease, might wish to take the test to see if they did, in fact, carry the faulty gene. Those who favoured testing pointed out that it would allow those carrying the disease gene to plan their lives and to avoid passing on the disease to their children. But once linkage testing became available, studies of programmes suggested that probably little more than 5 per cent of those at 50 per cent risk have chosen to be tested (Quaid & Morris 1993, Craufurd et al. 1989). Of those being tested, relatively few seem to wish to use testing for reproductive choice. Exclusion testing, which identifies a foetus at 50 per cent risk but avoids revealing the genetic situation of the parent, has also had a relatively low uptake (Craufurd & Harris 1989).

With hindsight, it is perhaps not too difficult to see why many family members do not wish to have linkage testing. While there is a 50 per cent chance of learning that they are very unlikely to get the disease, it is equally likely that they will learn that it is almost certain that they will get the disease (see Tibbens, Niermeijer et al. 1992). Most family members will know in all too painful detail what getting the disease will mean for them, having seen affected relatives. Hope for the future may lie in retaining the uncertainty, the fifty-fifty chance that they will not get the disease.

Now that direct testing is available, the high or low probability given by linkage testing can be replaced with a result which, in almost all cases, is near certain. In addition, direct testing can provide a result for someone who does not wish to involve other family members in the testing procedure, or who does not have enough family members able or willing to provide DNA samples for linkage testing. Whether or not these new factors will lead to different patterns of test uptake is as yet unclear. While there have been reports of an increase in uptake of testing (Rosser et al. 1994), it remains the case that the great majority of those who know themselves to be at risk of developing the disease have not come forward for testing.

There have been a number of studies of the consequences for those who have opted for linkage testing. Those who have been found to have a high risk of carrying the gene may experience psychological and emotional difficulties in coping with the information, but at least one study found that the initial shock was followed by a fairly rapid recovery (Tibbens, van de Vilis et al. 1992). Among those who were found to be at low risk, some have become very distressed (Huggins et al. 1992).

Others found the relief that they experienced on learning that they are unlikely to carry the gene was soon replaced by persistent guilt feelings and depression (Tibbens, van de Vilis et al. 1992). As well as these research studies, there are personal accounts, referred to above (in Marteau & Richards 1996).

Most of the individuals who have been tested are likely to have known about the disease and their own risk of getting it from an early age. They have grown up with the knowledge of uncertainty. It is on that basis that we have to understand reactions to test results. Having lived a long time with the knowledge of an uncertain future, this may have become part of a personal identity and it may be difficult to accept a test result. Additionally, when a result is negative, the guilt and depression that are sometimes reported may be akin to that sometimes seen in the survivors of other potentially traumatic situations. Another reason for being upset after receiving a negative test result is that it may lead to regret about earlier decisions, such as relationships not entered into or children not conceived.

But it must be emphasised that, despite almost ten years of linkage testing, we still lack adequate research evidence. While a few studies of reproductive choice have been carried out, they are very much in the mode of examining how far those at risk have received genetic counselling and the extent to which such counselling has prevented births of affected children (e.g. Tyler & Harper 1983). While studies such as this (see also Wexler 1979, Wexler 1995) have speculated about why 'many HC [Huntington's chorea] families continue to bear children in the full knowledge of the genetic risk', they provide little insight into the beliefs held about the disorder and the ways in which family members reach decisions about such matters as marriage and child-bearing. These approaches often assume that families need professional intervention to function adequately: 'the burden of telling children the risk was too great for most people and professional help was needed'.

Let me leave the last word with a member of a family with Huntington's disease. The following statement was written by a mother of four, who learned as a teenager that a grandparent had the disease and was told of the implications for herself and her parents. Her partner learned the implications very early in their relationship, and they had their first child after both her brother and sister had had children. The diagnosis of one of her parents came soon after the birth of their second child.

Looking back, it seems quite clear that I have spent over 20 years taking a 'head in the sand' approach to the whole question of Huntington's Disease – avoiding thinking about it, avoiding focusing on it, and somehow denying that it really has anything to do with me. Since telling two or three good friends when I was in my teens, I have avoided telling any friends, or indeed my partner's family, about the disease, and continue to do so. Until recently I have also sought relatively little information about the disease, and have avoided any situation of being identified as 'at risk', for instance by becoming a member of the Huntington's Disease Association – indeed, I have never considered seeing a genetic counsellor, and would strenuously resist putting myself in a position where my 'at risk' status would be entered in my medical records. That I have been able to sustain this way of dealing with the situation has perhaps been facilitated by the fact that there has so far been no question of being tested for the Huntington's gene, or of having prenatal diagnostic testing, since one of my parents is the only living member of the family with the disease, and testing using markers is therefore not possible. I do wonder whether my partner and I would have made different decisions had testing been available to me.

I suspect that I might have found the prospect of aborting an affected foetus unacceptable, and especially if this would have involved a relatively late abortion following amniocentesis (rather than an earlier one, following chorionic villus sampling). Yet, had testing been available, I would suspect I would also have felt it unacceptable to bring a potentially affected baby into the world when I could have avoided this with certainty. There would also have been the issue of whether I could have faced the possibility of a foetal test proving positive, and therefore showing that I definitely had the gene myself. I feel fairly sure that I would not have taken the option of exclusion testing, finding the possibility of aborting a foetus with only a 50 per cent chance of having the gene altogether too cavalier. (Interestingly, my sister recently told me that had this test been available to her, she might well have used it – her fear of finding out her own genetic status, combined with a profound desire not to pass the Huntington's gene on to another generation, making the possibility of aborting an unaffected foetus worth the risk). It does seem possible to me that, had testing been available, the only option I would have been able to contemplate with any equanimity would have been the one to avoid having children at all.

This, however, remains pure speculation. As regards the future, I am now aware that, the Huntington's gene itself having been located, a test will soon be available to me. There are days when I think that the obvious course is to have the test. I feel strongly that I do not want to spend the next 20 years worrying about something needlessly. On the other hand, should the test prove positive, my partner and I could plan our lives on the basis that my ability to earn an income was unreliable in the short-to-medium term, and would almost certainly not extend beyond my 50s, and on the basis that we should enjoy the present because we would not have much of a future together. On other days these arguments seem to have little validity – after all, no couple can assume they will be able to spend their retirement together, and it must be good advice to anybody to make the most of the present rather than always looking to the future for good times. In addition, it would clearly be a mistake to pass any opportunity by on the basis that I had the gene, and could develop the disease at any time. If I lived my life on this premise, I might find that by the time the disease struck advances had been made towards treatment, so I had let the disease limit my life unnecessarily. Were genetic testing able to reveal at what age one was liable to succumb to the disease, or were treatment available, the question of whether to be tested might be more clear cut. As things stand it is not. (Anon 1996)

HEREDITARY BREAST AND OVARIAN CANCER

While it has been known for some time that both breast and ovarian cancer can be familial (that is to say they may tend to cluster in families but not necessarily for genetic reasons), it has recently become clear that a small proportion of cases of each (probably around 5 per cent) are related to one of probably a small number of highly penetrant dominant gene mutations. The most important of these, which has been called BRCA1, has been cloned and is probably associated with at least half the cases of hereditary breast and ovarian cancer. The association is most frequent when the onset of the cancer occurs at a particularly young age (Evans et al. 1994). Because the faulty gene has been located and markers are available, linkage testing can now be done in the same way as for Huntington's disease for some families (Biesecker et al. 1993, Evans et al. 1994) and direct testing is available for a few families with a known mutation.

But there are some very important differences between the situation for Huntington's disease and that for breast and ovarian cancer. While Huntington's disease is 'purely genetic' – that is, it does not occur without the presence of the faulty gene – most cases (95 per cent) of breast and ovarian cancer are not linked to a faulty gene in this way. While cases linked to BRCA1 tend to occur earlier in life, the form of the disease itself is not different in these cases and the disease cannot be regarded as 'genetic' in the same sense as Huntington's disease.

The Huntington's disease mutation is 100 per cent penetrant: that is to say that any individual who has the faulty gene will develop the disease providing that they live long enough. BRCA1 is probably about 80 per cent penetrant, so that about 20 per cent of those who carry the gene mutations will not get breast or ovarian cancer (the gene is not expressed, to use the geneticists' terminology). Both the Huntington's disease gene and the BRCA1 gene are dominantly inherited, but in the case of the latter, the cancers linked to it only occur in women (except in very rare cases of male breast cancer which may be linked to BRCA2). Thus, men may carry a faulty BRCA1 gene and pass it on to their children, but they themselves will not get breast or ovarian cancer. They are normally healthy (though in a few rare cases they may develop other cancers apparently associated with the gene). So, because the penetrance of BRCA1 is less than 100 per cent in women, and because the cancers are not seen in men, the pattern of the occurrence of cancer in a family who carry BRCA1 is more intermittent than the pattern of occurrence in a family with Huntington's disease.

The other important respect in which the two diseases differ is that breast and ovarian cancer are much more common in the population than Huntington's disease. While Huntington's disease is thought to have an incidence of one in 5000 at birth, the lifetime risk for breast cancer is about one in 12 women and that for ovarian cancer is one in 120 women. Thus, breast cancer, in particular, is very much part of the common experience of the population and is widely discussed in the media. There is now a national breast screening programme for women aged 50 or more who are offered regular mammography. There are also many schemes to teach techniques of breast self-examination. At present, effective methods for population screening for ovarian cancer are not available.

Our research with families with hereditary breast and ovarian cancer, which is still at an early stage, involves women who come to a clinic for family cancer syndromes. At this, they can receive information about the likelihood of their carrying a gene mutation which predisposes them to breast or ovarian cancer, and advice is available on what they might do about the predisposition through screening or possibly prophylactic surgery (see Richards et al. 1995). The basic technique used in these clinics is to collect a detailed family history over at least three generations which can then be used to calculate an individual's risk of carrying BRCA1 (or a similar gene) and so their chances of developing a cancer. In a few cases, where there is an appropriate family history and the relevant family members agree, linkage testing may be offered. This can indicate whether an individual has a high or low probability of carrying BRCA1. Because BRCA1 has only recently been cloned and is a large and complex gene with many mutations, direct testing is not generally available, although some direct testing is being carried out in families with a known mutation on a very small scale and on an experimental basis.

For our research project, we interview women before their clinic sessions and again a few weeks later. We tape-record the clinic sessions. The major aim of these interviews is to explore knowledge about the inheritance of breast and ovarian cancer as well as other characteristics, to investigate communication within their families about such matters, and to describe both how women cope with the possibility of developing a cancer and their family situation in general.

Most women currently coming to the clinic are self-referred. They have usually been aware for some time that cancer 'runs in their family'. The main motive for coming to the clinic is to obtain advice on early detection of breast and ovarian cancer and other actions they could take and, in a few cases, about prophylactic surgery. Most seem less interested in receiving information about their precise genetic risk. Given that they already assume that they may be particularly prone to the disease in most cases, it is understandable that their main interest is in ways of dealing with the potential threat to themselves and, in some cases, to their children.

A typical history is that the women become aware of breast or ovarian cancer in their family as a teenager or young woman: 'I remember [my mother] telling me when her mother developed breast cancer. I was about nine or ten at the time.' They have been prompted to come to the clinic, or have heard about it, through press and media discussion of breast and ovarian cancer genetics. For others, the prompt to come has been the recent death from breast or ovarian cancer of a close relative or the realisation that they are approaching the age when relatives have died from one of the diseases.

While most of those coming to the clinic know that breast and/or ovarian cancer may be inherited and runs in their families (ideas which are likely to be reinforced by recent press coverage), their understanding of the process is seldom Mendelian. They usually see the 'proneness' to these cancers as something that is passed through the family, often from grandmother to mother and so to daughters (see Green et al. 1993). Few family histories correspond to such a regular pattern, but there are a number of common beliefs that appear to serve the function of reconciling this general assumption to particular patterns seen in a family.

The most widespread of these is that the diseases may 'skip a generation', an idea that can be traced back at least to Aristotle's ideas about inheritance (see Russell 1986). It is not a concept that appears to be strictly applied, simply a belief that the appearance of the disorder in the family tree will be irregular.

Another recurring idea that appears to serve a similar function is that there will be a resemblance, physical or in character, between those in the family who do or do not get the disease. So, for example, one woman felt her risk of getting the disease was very high as she resembled her mother in many ways, while her sister resembled an aunt, 'who is still going strong at a ripe old age, so she [the sister] will be in the clear'.

An idea that many hold (which is similar to that of some geneticists) is that a person is likely to develop the cancer at the same age as it developed in their nearest relative who had the disease. This notion seems particularly likely to be held by daughters in relation to their mothers. Occasionally this leads a family member to believe that they can't be 'prone' to the disorder as they have already passed what they regard as the age at which members of their family get the cancer.

Family members often find it hard to understand that the gene can be passed on via a male: 'I don't think the gene is carried by the male line, is it? It's the female line I suppose' (A woman three weeks after counselling). This is sometimes true even in families where the distribution of cases makes it very likely that this must have happened. It is interesting to note that a bias may also be seen in some pedigrees produced by professionals who do not always pursue male branches of the family as thoroughly as female. (The same effect may be evident in the paper by Evans et al. (1994), where the situation of offspring of a male who is at a 50 per cent chance of inheriting the disorder is not mentioned.) An even more striking example of the extent to which breast cancer is seen as only transmitted via females is of a man who developed breast cancer and who had three first-degree female relatives who also had breast cancer. After genetic counselling this man immediately informed his daughters of their potential risk, but he had simply told his two sons (one of whom had daughters) about his own experience of breast cancer without pointing out to them their own potential risk or that of their children.

The difficulty of appreciating inheritance via males may simply be a difficulty in associating men with diseases which are almost entirely female. It could also arise because of the ideas held about the process of inheritance, which will be discussed further below. Many seem to believe in a propensity or proneness to the cancer itself which passes directly from one family member to another, rather than a segregating gene that may be expressed. If a woman has this proneness she is likely to get the disorder. If she has not inherited the proneness, she is no more likely to get the disease than others who do not come from a cancer prone family. Opinions appear to be divided on whether or not other environmental and lifestyle predisposing factors may influence the inherited proneness or whether the two work independently. One woman thought she might not get breast cancer, despite coming from a cancer family, because she did not smoke. Another felt her sister to be particularly at risk because she was always getting ill and was not a particularly healthy person. Others saw the two kinds of risk operating independently, and they felt there was nothing they could do to alter their chances of getting the disorder. Some of those who adopted this position believed that the cases of cancer in their family could be divided into inherited and environmental cases. (This will be true sometimes – as women in these families who do not carry the faulty gene are still at risk of developing a cancer like anyone else in the population.)

At the beginning of the counselling session, women were usually asked whether or not they saw themselves to be at risk of developing whichever of the two cancers (or both) had occurred in their family.

Almost all said they did and they would often put their risk very high (see also Evans et al. 1993). A few women believed they were certain to develop a cancer – 'just like everyone else in the family' – and the only question was when. One woman who was certain about this came to discuss treatment options while this could be done without the pressures of time she felt would be present once the diagnosis had been made. It, of course, follows that, while only 50 per cent of those who are a child of someone with a BRCA1 (or 2) mutation will inherit it, and only in a proportion of these cases will the mutation be expressed, risks understood on the 'proneness' model can be seen to run as high as 100 per cent.

Many women are aware before the session that some kind of test for the gene may be soon available. This future possibility of direct testing is usually discussed in the session and they are given a rather cautious assessment of when it might be available. It is pointed out to them that such tests will simply refine the estimate of their risk of familial breast cancer, rather than telling them whether or not they will be affected (because BRCA1 is not 100 per cent penetrant). From interviews carried out after the counselling, it is clear that some, at least, believe that if such a test is offered, it will only be so if and when effective treatments for the disease are also available.

I would assume from the knowledge so far that if they were prepared to offer me the test, they would then be prepared to come up with a treatment that would either contain it, kill it or monitor it so that it didn't get to the stage of killing me. So, yes, I would do it [take a genetic test]. (A woman interviewed three weeks after counselling)

If views such as this are widespread, it is perhaps not surprising that many women say that they would be interested in a test which could determine whether or not they carry the gene. Also, it is not widely understood that the test will only be informative in families that carry specific BRCA1 mutations, rather than in families with other dominant genes related to the cancers.

While some family members are specifically concerned with the potential risks for their children, like the (medically qualified) father who came for information about his teenage daughter after the death of his wife from breast cancer, most of those who come do not have issues related to their children uppermost among their concerns. Like Evans et al. (1994), we find that most women who come have already completed their child-bearing and want primarily to discuss their own situation. In several sessions, the issue of children's possible risks was first raised by the counsellor. When daughters may be at increased risk, most mothers suggest they will tell them as young adults, often expressing the hope that, by then, 'something' could be done for them. Less often, such concerns are volunteered about sons. Most undertake to inform other relevant members of the family about the counselling and how the situation may affect them (see Green et al. (in press) for further discussion of this issue). In a few cases, written material (including the letters summarising the outcomes of a session from the counsellor) is passed on to all family members. Because of the need to collect further information about the family history, it is not uncommon for links to be renewed with family members who had not been in touch for several years.

It is important to note that not all members of the same family will take the same approach to dealing with the threat of inherited breast or ovarian cancer. We have already found a number of families where two sisters seem to cope with it in radically different ways – one seeking out information, wanting to know as much as possible, while the other avoids the topic and may describe herself as being very reluctant to come to counselling.

Discussion of possible screening for cancer takes up a significant part of the counselling sessions. At least for women who are at above population risk, all potential methods of screening are discussed, often in considerable detail. These discussions usually include information about the effectiveness of techniques and their shortcomings. Women are usually given direct advice about what might be their best strategy, given the estimate of their risk. Most women seem to expect direct advice. The little follow-up data we have suggests that part of the satisfaction with the counselling (expressed sometimes even after they had been given a risk estimate higher than expected) was that they had been given direct advice about what to do. In this sense, the counselling is directive. Most women have heard about cases of prophylactic mastectomy and oophorectomy. While the latter was advised for women at raised risk of ovarian cancer who have completed their child-bearing and show any interest in this approach, a much more conservative view is taken of mastectomy. Even women who are probably at a 50 per cent chance of carrying the BRCA1 gene are advised that this is a very serious step that they should consider very carefully. Some women indicated that they wanted to wait for the advent of direct testing before considering this.

As yet, we still have to complete follow-up interviews after counselling sessions. But such information as we have suggests that the information they have been given about risk cannot always be repeated, and the understanding of the genetic explanation offered in the clinic is often limited.

SOME HYPOTHESES ABOUT LAY KNOWLEDGE OF INHERITANCE AND KINSHIP

Our research, both with members of families who carry a genetic disorder and those who do not, suggests that it is rare to find people who describe inheritance in terms of the scientific account. If they do, it is usually because they have had specific training in biology and genetics beyond school level. Those who have received genetic counselling, which usually includes an outline of the process of inheritance in Mendelian terms, are seldom able to reproduce this account if asked subsequently. These preliminary findings would be consistent with research in the public understanding of science tradition (Durant et al. 1996). However, it should not be assumed from this that there is not a widespread lay knowledge of inheritance. Indeed, everyday observation of English family life suggests that there is a general interest in issues of inheritance. Family discussions of a newborn baby or of family photographs or portraits centre on questions of resemblance and who takes after who. Different features may be assigned to different family members: 'He has his father's eyes but his mother's nose.' Or a feature may be seen as a property of one side of the family: 'He has the Jones' temper.' As these examples indicate, features that are assumed to be inherited may include both physical attributes and aspects of behaviour and character. When questioned about particular characteristics, there seem to be fairly consistent beliefs about which are largely a matter of inheritance (for instance, eye colour, baldness and body build) and which are more a matter of nurture (especially psychological characteristics). Similar consistent patterns may be found among diseases, with heart disease, for instance, being seen as largely inherited (Davison et al. 1989, Ponder et al. 1996).

The sense of inherited similarities runs parallel with, and indeed may be seen as part of, our conceptualisations of kinship (see Strathern 1992). The English kinship system is bilateral so that there is a general view that a more or less equal contribution for a child will be derived from each parent. An exception to this may be sexual or sex linked characteristics. This seems to be the case with breast cancer, where many have difficulty with the idea that this may be inherited via a male member of the family (as has already been mentioned).

Inheritance is generally seen in terms of a biological connection. While kinship and biological connection are not synonymous, they are clearly closely linked (Richards 1996). 'Real' fathers, unlike step-fathers, are biological parents (or assumed to be biological fathers). The discovery that a child has not been fathered biologically by a parent can lead to an ending of the social relationship with that parent (Cohen 1995). However, it is possible to find examples where it seems that a kinship link which does not involve a biological link (such as a step-parent) is seen to have led to inheritance of a characteristic. Here we should note that the term inheritance has a dual meaning – the passing on of characteristics through a process of biological transmission, as well as the handing down through a family of property after death.

The degree to which biological inheritance is shared within a family appears to parallel ideas of closeness of kinship, though kinship is a social construction. In formal terms it may be defined as a set of beliefs, values and categories which structure relationships and social perceptions and actions (Harris 1990). English kinship is egocentric and defined in relation to the individual. While we talk of our parents and children, we are also our children's parents (see Sneider 1968). Our closest kin are those with whom we are likely to reside for a part of our life and have intimate emotional relations, and for whom we are seen to have responsibilities (i.e. our parents and children). And, indeed, these are the kin with whom we see close links of (biological) inheritance. According to some preliminary work, when people are asked about the degree of shared inheritance with relatives, most seem to be able to express this accurately (in a genetic sense) when referring to their parents and children but are much less certain about sisters and uncles and appear to underestimate the genetic connection with these two latter categories (Richards & Ponder 1996).

Inherited connections within a family are often discussed in terms of blood relationships, a term that can be traced back to an Aristotelian view of inheritance. Indeed, there is a common belief that blood donations from those who carry a genetic disease may present dangers for the recipients. Anthropological accounts describe kinship in these terms:

the blood that flows between English relatives...is at once like a moving stream (and cannot travel backwards) and like a substance that can be infinitely divided into parts. Hence the 'dilution' of any one stream that comes from mixing, rendering any one individual as an amalgam of blood...in popular belief, the parts that an individual person 'gets' from either mother or father may be thought of as parts of other ancestors that 'show' in descending generations. (Strathern 1992: 80)

Despite these ideas about blood, it is not generally believed that blood is the substance that is contributed by each parent to a child at conception. Almost universally, adult and some child informants agree that conception occurs through the union of an egg and sperm. However, lay accounts do not seem to specify how inherited features are passed on. What is clear is that the century-old scientific account of a reduction division of genetic material in the formation of eggs and sperm, and their union in the zygote with the fusion of the two sets of chromosomes, is not generally reflected in lay knowledge. There appears to be no concept of a genetic material independent of other body substances (such as blood), nor of the precise formulation of the mode of transmission of the inherited influence. It would seem that, in lay knowledge, there are separate domains concerned with sexual intercourse and conception on the one hand, and with the transmission of inherited characteristics on the other. Given this, it is not surprising that many find a scientific account of segregating Mendelian genes provided in the genetics clinic (or at school – see Ponder et al. 1995) very difficult to understand. Terms like 'gene', 'DNA' or 'chromosome' are used widely in popular discourse (Nelkin & Lindee 1995), but this should not be taken to mean that the users have a scientific knowledge of inheritance. Rather, in this context, 'gene' seems to be a term for the general concept of the biological transmission of characteristics between generations. The varied usage of such terms by different social groups and in different contexts presents obvious complications for researchers interested in knowledge of inheritance.

Given that lay knowledge does not seem to embody a concept of a specific substance which is responsible for passing on inherited characteristics between generations, it is not surprising that many people talk of the direct transmission of the characteristic (such as breast cancer), or a 'proneness' to it, between parent and child (see Green et al. 1993). There is no space in such conceptual schemes for varying expressions of a gene. Therefore, in the context of genetics clinics, it may be very difficult to explain the inheritance of conditions that are less than 100 per cent penetrant or the inheritance of a 'silent' gene as the male transmission of a BRCA1 mutation.

It seems reasonable to assume that issues of inheritance will be far more salient in families that carry a genetic condition than for the rest of the population. As yet, however, too little work has been done with non-clinical populations to be clear about this. What does seem possible is that some people who have received genetic counselling may see the inheritance of the specific condition that was the subject of the counselling in Mendelian terms, but that their more general notion of inheritance may remain unaltered in a separate domain which they continue to use in discussing such matters as family resemblances. Perhaps this is what happens when school children are taught genetics as part of their science curriculum; a new domain of scientific knowledge is produced which exists quite independently of everyday ideas about inheritance and family resemblances. And it may be that such knowledge is effectively unavailable should that child later in life need to confront issues related to genetic disease.

CONCLUSIONS

This paper has been intentionally speculative. It is intended to provide signposts in an area we need to explore if we are to understand family members' knowledge of genetic disorders they may carry, and hence the choices they make in connection with such family matters as reproduction, as well as the use they may, or may not, make of genetics services. If the general perspective I have outlined is broadly correct, there are obvious implications for genetic counselling as well as genetic education. Explanations of an individual's genetic situation may best proceed from an understanding of that individual's prior knowledge of inheritance. There are also implications for more general education in Mendelian and molecular genetics in school and elsewhere (these issues are pursued further in Richards 1996, Richards & Ponder 1996).

When linkage testing for Huntington's disease was on the horizon, its potential uptake was greatly overestimated. There are several reasons for this, including people's need to hold onto a belief that they might not get an incurable and devastating disease. This example also underlines the point that decisions about reproduction and other issues of daily life are a great deal more complex than a series of rational choices made in the light of scientific information about a disease and its inheritance. Indeed, the whole incident underlines a pressing need for much more wide-ranging studies of reproductive choice in couples who carry genetic disease.

In the case of BRCA1, linkage testing is now underway. Here it seems likely that the response will be very different from that towards testing for Huntington's disease. Indeed, the major difficulty seems likely to be to persuade women that direct BRCA1 mutation testing is only likely to provide useful information for a very small group of people.

The use of genetics services is very much a cultural matter, and what I have said about testing for Huntington's disease and BRCA1 may well not hold good outside the English culture where our research has been carried out.

ACKNOWLEDGEMENTS

The research on families and genetic disorders is carried out in collaboration with Josephine Green, Helen Statham, Frances Murton and Nina Hallowell and is supported by a grant from the MRC. We are grateful to Professor Bruce Ponder and some of his patients for their collaboration in our research

REFERENCES

Anon. 1996 Living with the threat of Huntington's disease, in Marteau, T., & Richards, M. (eds) *The Troubled Helix: Social and psychological implications of the new human genetics.* Cambridge University Press: Cambridge.
Biesecker, B.B., Boehinke, M., Calzone, K., Markel, D.S. et al. 1993 Genetic counselling for families with inherited susceptibility to breast and ovarian cancer. *Journal of American Medical Association* 269: 1970–4.
Cohen, D. 1995 So you think you know. *The Independent*, 21 July.
Craufurd, D., Dodge, A., Kerzin-Storrar, L., & Harris, R. 1989 Uptake of presymptomatic predictive testing for Huntington's disease. *Lancet* II: 603–5.
Craufurd, D., & Harris, R. 1989 Predictive testing for Huntington's disease. *BMJ* 298: 892.
Davison, C., Frankel, S., & Smith, G.D.I. 1989 Inheriting heart trouble: the relevance of common sense ideas to preventive measures. *Health Education Research* 4: 329–40.
Durant, J., Hansen, A., & Bower, M. 1996 Public understanding of the new genetics, in Marteau, T., & Richards, M. (eds) *The Troubled Helix: Social*

and psychological implications of the new human genetics. Cambridge University Press: Cambridge.

Evans, D.G.R., Burnell, L.D., Hopwood, P., & Howell, A. 1993 Perception of risk in women with a family history of breast cancer. *British Journal of Cancer* 67: 612–14.

Evans, D.G.R., Fentiman, I.S., McPherson, K., Asbury, D. et al. 1994 Familial breast cancer. *BMJ* 308: 183–7.

Green, J.M., Murton, F., & Statham, H. 1993 Psychosocial issues raised by a familial ovarian cancer register. *Journal of Medical Genetics* 20: 575–9.

Green, J.M., Richards, M., Murton, F., Statham, H., & Hallowell, N. In press. Family communication and genetic counselling: the case of hereditary breast and ovarian cancer. *Journal of Genetic Counselling.*

Harris, C.C. 1990 *Kinship.* Open University Press: Milton Keynes.

Huggins, M., Bloch, M., Wiggins, S., Adams, S. et al. 1992 Predictive testing for Huntington's disease in Canada: adverse effects and unexpected results in those receiving a decreased risk. *American Journal of Medical Genetics* 42: 508–15.

Huntington's Disease Collaborative Research Group 1993 A novel gene containing a trinucleotide repeat that is expanded and unstable on Huntington's disease chromosome. *Cell* 72: 971–83.

Klein, J. 1982 *Woody Guthrie: A life.* Ballantine Books: New York.

Marteau, T., & Richards, M. (eds) 1996 *The Troubled Helix: Social and psychological implications of the new human genetics.* Cambridge University Press: Cambridge.

Nelkin, D., & Lindee, M.S. 1995 *The DNA Mystique: The gene as a cultural icon.* Freeman: New York.

Ponder, M., Lee, J., Green, J.M., & Richards, M. 1996 Family history and perceived vulnerability to some common diseases: a study of young people and their parents. *Journal of Medical Genetics* 33: 485–92.

Quaid, K.A., & Morris, M. 1993 Reluctance to undergo predictive testing: the case of Huntington's disease. *American Journal of Medical Genetics* 45: 41–5.

Richards, M. 1993 The new genetics: some issues for social scientists. *Sociology of Health and Illness* 15: 567–86.

Richards, M. 1996 Lay and professional knowledge of genetics and inheritance. *Public Understanding of Science* 5: 217–30.

Richards, M., & Green, J.M. 1993 Attitudes towards prenatal screening of fetal abnormality and detection of carriers of genetic disease: a discussion paper. *Journal of Reproductive and Infant Psychology* 11: 49–56.

Richards, M., Hallowell, N., Green, J.M., Murton, F., & Statham, H. 1995 Counselling families with hereditary breast and ovarian cancer: a psychosocial perspective. *Journal of Genetic Counselling* 4: 219–33.

Richards, M., & Ponder, M. 1996 Lay understanding of genetics: a test of a hypothesis. *Journal of Medical Genetics* 33: 1032–6.

Rosser, E., Huson, S.M., & Norbury, G. 1994 Prenatal, presymptomatic and diagnostic testing with direct mutation analysis in Huntington's disease. *Lancet* 343: 487–8.

Russell, N. 1986 *Like Engendering Like: Heredity and animal breeding in early modern England.* Cambridge University Press: Cambridge.

Sneider, D.M. 1968 *American Kinship: A cultural account.* Prentice Hall: Englewood Cliffs.

Strathern, M. 1992 *After Nature: English kinship in the late twentieth century.* Cambridge University Press: Cambridge.

Tibbens, A., Niermeijer, M.F., Roos, R.A.C. et al. 1992 Understanding the low uptake of presymptomatic DNA testing for Huntington's disease. *Lancet* 340: 1416.

Tibbens, A., van de Vilis, M.V., Skraasted, M.I. et al. 1992 DNA testing for Huntington's disease in the Netherlands: a retrospective study on psychological effects. *American Journal of Medical Genetics* 44: 94–9.

Tyler, A., & Harper, P.S. 1983 Attitude of subjects at risk and their relation toward genetic counselling in Huntington's chorea. *Journal of Medical Genetics* 20: 179–88.

Weatherall, D.J. 1991 *The New Genetics and Clinical Practice*, 3rd edition. Oxford University Press: Oxford.

Wexler, A. 1995 *Mapping Fate: A memoir of family risk and genetic research.* University of California Press: Berkeley, Calif.

Wexler, N.S. 1979 Genetic Russian roulette: the experience of being at risk for Huntington's disease, in Kessler S. (ed.) *Genetic Counselling: Psychological dimensions.* Academic Press: New York.

Wexler, N.S. 1992 Clairvoyance and caution: repercussions from the Human Genome Project, in Kevles, D.J., & Hood, L. (eds) *The Code of Codes: Scientific and social issues in the Human Genome Project.* Harvard University Press: Cambridge, Mass.

Part IV
Social and Political Issues

14 'Once You Have a Hammer, Everything Looks Like a Nail'

Pat Spallone

> She could not understand for she did not know the pattern of the ethics used in the hospital way of doing things, but she had the faith in me to see that I must be acting according to a pattern. How was she to know the ethics of relationships that science and the machine were demanding of the world, that had their roots in other continents?
>
> Ida Pruitt,
> *Old Madam Yin: A memoir of Beijing life 1926–1938*

The report in 1993 that scientists had located a gene linked to homosexuality is only one in a growing list of 'finds' over the last few years associating genes with diseases and non-disease states. My personal favourite was the headline, 'Faulty genes lead to old age'.[1]

The dramatic increase in the number and range of genetic tests designed to screen embryos, foetuses, children and adults for inherited or congenital disorders, and for predispositions to complex conditions such as cancers or asbestosis (which is caused by exposure to asbestos especially in work settings) all point to the expanding scope of genetic screening.

A number of areas of concern have been identified with the expansion of medical genetics, and they have become almost common sense. Concrete proposals have been forthcoming to address these concerns. In this chapter, I will introduce some of these proposals, and then use them as a springboard to reflect on some of the 'ethics of relationships' which are implicit in them. This phrase has been borrowed from the medical social worker Ida Pruitt, whose understanding of things Chinese was rivalled only by her understanding of things Western.

THREE AREAS OF CONCERN

Three key areas of concern which have become common sense, and which I shall briefly discuss, are the possibilities of genetic discrimination, the issue of informed consent, and the consequences of the new genetics in relation to medical practice.

1. Genetic discrimination as a civil rights issue
The possibility of genetic discrimination arises, as does a new concept dubbed 'the healthy ill'. Individuals labelled as predisposed to illnesses may become a new group to be discriminated against as a category of 'at risk' people. They may be denied insurance or employment, as has already occurred in the USA (Billings 1991). Allied to this is the question, 'What is the impact of expanding screening on the way society views and supports disabled people?'

2. Informed consent and choice
A second set of concerns revolves around the issue of informed consent: ensuring that individuals are given the opportunity to make meaningful choices in screening programmes.

The late Wendy Farrant's important early study (1985) of the views of consultants and the experiences of women of prenatal diagnosis raised several points which are still relevant. Farrant found a view among consultants that selective abortion is desirable when foetuses are diagnosed as disabled; and that at the various stages of testing, counselling is biased towards encouraging women to take up the tests and have an abortion if abnormality is detected. Thus, the professional ethos – that women should make their own decisions – was contradicted in an underlying assumption that the decision to enter a screening programme is tantamount to a decision to have an abortion if abnormality is detected.

Farrant also found that more than 25 per cent of the consultants surveyed routinely administered AFP screening (alpha-fetoprotein tests, which can show up neural tube defects) without offering the woman any explanation, and although the test was supposed to be voluntary.

As others observing similar findings point out, judgements and assumptions about impairment among consultants may reflect biased views about impairment in society at large (Marteau 1994, Green 1990, 1995)

3. Genetic priorities
The most far-reaching consequence of the new genetics in clinical practice may be a changing medical model which takes its cue from genetic capabilities and ideas. The Council for Responsible Genetics (1990) in the USA – supported by many scientists – stated in their position paper on genetic discrimination:

> The exaggerated emphasis on genetic diagnosis is not without its dangers because it draws attention away from the social measures which are needed in order to ameliorate most diseases, including equitable access to health care. Once socially stigmatised behaviours, such as alcoholism or other forms of addiction or mental illness become included under the umbrella of 'genetic diseases', economic and social resources are likely to be diverted into finding biomedical 'cures', while social measures will be short-changed. (p. 288)

Once you have a hammer, everything looks like a nail. As I finalise this text, fresh controversy surrounds claims of evidence of genetic determinants for intelligence and aggression (Radford 1995). The tools of science, as Stacey suggests in Chapter 17 of this volume, must not be embraced without critical awareness.

SOME PROPOSALS

How then may we safeguard against the more immediate problems of protection against discrimination, and protection of the individual's right to consent as an informed, willing and uncoerced commitment to a decision? Various proposals have been put forward:

1. Develop fact sheets that describe what is known about genetic screening, and why genetic status does not necessarily identify an individual's health or abilities.
2. Offer short courses on the uses of genetic screening to professionals and the general public, preferably together so that neither group gets stuck in its own assumptions.
3. Draft model laws which prohibit discrimination based on present or predicted medical status and hereditary traits.
4. Propose legally binding agreements on confidentiality to protect information obtained from genetic screening.

5. Recognise the need for systematic social science data on the implications of medical genetics research and practice, and the voices of
the consumers and 'the public', to inform a wide debate.

I support such proposals, but certain questions arise: for example, 'How
is information, such as the facts in fact sheets, communicated?' A prevailing model is the deficit model of the public understanding of science,
which assumes that there is a deficit of knowledge about the scientific
facts (in this case, the facts of genetics) among the general public. Once
educated in the facts, the public and society will be in a better position to
judge and utilise developments. Generally implicit in this model is an
assumption that scientific facts are mirrors of nature, and everything else
is opinion, values, ethics and so on. For a concrete example, consultation documents on controversial developments in reproductive and
genetic medicine assert that while it is valid for scientists to discover as
much as possible about biological systems, 'science must have an ethical
context' (HFEA 1994: 2). The deficit model of the public understanding
of science bifurcates knowledge and the ethical context, allowing that
attention to ethics will safeguard against abuse.

This model, even at its best, is limited because it cannot engage with
the ethics of relationships demanded by the science. This can be
illustrated with two examples.

Ethics of Relationships

1   The average foetus: where an ethics of relationships is implied
Historian Barbara Duden tells of a visit to a prenatal centre in New
York's Harlem area, where people from less privileged parts of the
Caribbean are pouring in. One pregnant woman who can hardly
understand English is being interviewed, empathetically, and is being
told of the greater risk to the foetus now that she is over 35 years of
age. This is the woman's sixth pregnancy, and her mother told her that
the older the mother, the brighter the child. Who could deny the
benefits to child and mother of the pregnancy? Duden points out
that the counsellor's insistence on risk made little sense to the
woman, yet from this point on, in all parts of her life, the woman is
to be 'bombarded with a dozen notions that together make up the
conceptual framework for a slum pregnancy in New York: normal
developments, risk, expectancy, fetus, social security payments, and
the like'. By contrast, the experience of her mother was more sensual,
warm, touchable, familiar. Duden concludes:

When this sort of question is raised in a medical milieu, it usually leads to a call for better preparation of the counsellor.... My point is that the procedures themselves must be questioned; not, however, from the perspective of their technical efficacy, but of their inevitable psychic results. (Duden 1993: 581)

The 'public understanding of science' model is not equipped to take into account the ethics of the relationships that science and the hospital way of doing things demand of this woman.

Similarly, in the concept of the healthy ill, mentioned above (a new category of at-risk person), there is an ethics of relationship being demanded, and it, too, has inevitable psychic results for the individual and for everyone.

2. Shared genes: where an explicit ethics of relationship is demanded
So far, I have been talking about an implicit ethics of relationships. I should like to offer a second scenario whereby the science explicitly sets an ethics of relationships.

The recent report of the Nuffield Council on Bioethics, *Genetic Screening: Ethical issues* (1993), and the earlier government report of the Committee on the Ethics of Gene Therapy (1992), both offered that patient confidentiality might be overridden in certain cases due to a concept of shared genes.[2] That is, if the patient refuses to release information about his or her genetic status to family members who may have an interest in the information, this exceptional circumstance might justify disclosure of the information by health professionals.

Here an ethics of relationships is being demanded. This ethic of *relatedness* is not totally new, of course. A similar ethical dilemma arises with HIV/Aids. But an ethic of relatedness also has positive connotations; it is an idea that feminist philosophers are exploring in relation to pregnancy, which has never fitted neatly into the political philosophy of individual citizenship and rights.

But why is relatedness an idea whose time has come now – that is, with the idea of shared genes and the techniques to identify them? On the one hand, you might say the idea has a self-evident power: the facts stand up for it. On the other hand, we might reasonably ask, 'How do ideas acquire power?'. Ludmilla Jordanova (1993), a historian of science, offers a theoretical insight. She, as others, argues that we are socialised into an essentialist way of thinking that focuses on the intrinsic properties of objects, a way of thinking profoundly akin to traditions of scientific thought. This way of thinking privileges things and events (in my example, genes and genetic events) over relationships and processes.

This insight does not make the ethical dilemma of confidentiality and the right to know go away, but it may help us to think about it and about other aspects of our subject with more understanding. It challenges the assumption that certain ideas are self-evident, and assumptions about what is common sense. It also exemplifies a point made by Sharma in Chapter 4, that we – whether as members of the public or the media, or scientists or other professionals – are all using some of the same concepts but not necessarily in the same way. This is a problem relevant not only to the concept of an ethnic group, but to concepts such as disabled person, at-risk individual, at-risk group, genetic cause, genetic solution and genetic counselling, too.

HOW DO IDEAS ACQUIRE POWER?

The question 'How do ideas acquire power?' may be raised in several contexts in relation to genetic disease in Britain, such as the context of consanguinity and genetic risk. Consanguinious marriage between first cousins within Asian communities in Britain is an issue which is often exaggerated out of all proportion to its real importance because it is founded on cultural prejudices embedded in Western medicine.

The question 'How do ideas acquire power?' is also important in relation to the themes of this chapter. As Marteau said when reflecting on the consequences of screening for Down's syndrome, 'Why do we have antenatal testing for Down's syndrome? The answer is: Because it is technically feasible. It is important to see how screening programmes are driven by technology' (Marteau 1994: 5).

It is also interesting to take account of a gendered view of these developments, as in the gendered consequence of Duden's story of the average foetus in Harlem; there is the question, for example, of the loss of other ways of knowing and experiencing pregnancy. (See, for example, Lippman 1994.)

Finally, I want to offer an altogether different example of what can happen when ideas acquire power. We need to be concerned about possible distortions of evidence due to investigators' attitudes and beliefs about genetics. J. Richard Marshall (1993) reviewed the report of the World Health Organisation's ten-country study of schizophrenia. The WHO study concludes that there must be a genetic basis of schizophrenia, and speculates on future use of genome mapping techniques – another multimillion-dollar world study. Marshall argues that the study's conclusions bear little relation to the evidence: a study which is likely to be accepted as a seminal publication produced conclusions which quite simply did not follow from their evidential premises. His point is how little substance there can be in what becomes accepted as conventional wisdom.

I would add that conventional wisdom based on faulty evidence also has psychic results, demands certain kinds of relationships, and is a dangerous consequence of the trend 'once you have a hammer, everything looks like a nail'. Social facts, attitudes, beliefs about heredity – for example, the consequences of consangineous marriage, the relationship between genes and behaviour, and so on – are usually seen simply as problems in the public understanding of science. But there comes a point at which all of us are the public, experts and laity alike, sharing some beliefs and values while differing in others.

SUMMARY AND CONCLUSION

My aim has been to identify areas of concern and concrete proposals which have arisen with expanding genetic medicine; but also to get past the 'common sense' of them, and take the opportunity to think about the complicated ethics of relationships where culture, kinship and genetics intersect.

The bifurcation of feelings and facts, emotions and logic, values and science, militates against a far-reaching and consistent analysis of reproductive and genetic technologies. We need to find more elegant ways of talking about these things, to paraphrase Marteau, than the prevailing 'public understanding of science' model allows.

For all of us, an adequate understanding of genetics requires a sensibility which integrates all of its aspects. In Chapter 11, Karbani and colleagues suggest that genetic information is never neutral, and that we need to listen to people's stories about their lives to expand our understanding of these matters. In Chapter 5, Darr tells of the success of communicating information from health studies through neighbourhood and community networks, rather than through typical health education campaigns, which are so often alarming and in which information is supplied in the form of stark facts and figures.

Another possibility may be to use the opportunity of information sheets and short courses (a strategy which I mentioned above) to challenge conventional images of disability, and to address prejudices and stereotypes of appearance and mental and physical capacities. Such an exercise could become part of the context in which genetic ideas are generated and played out. At least this approach begins to suggest a working plan which gives its due to culture – that is, to the interrelationships of the technical and social in our lives.

NOTES

1. Mohun 1987; see also Miller 1993, Spallone 1992: 175.
2. The relevant paragraphs are *Genetic Screening: Ethical issues*, 5.5–5.13; Committee on the Ethics of Gene Therapy, 4.15.

REFERENCES

Billings, P. 1991 Privacy violations arising from genetic discrimination. Testimony to US Congress Subcommittee on Government Information, Justice and Agriculture, with thanks to Paul Martin for personal communication.
Committee on the Ethics of Gene Therapy 1992 *Report* Cm 1788. HMSO: London.
Council for Responsible Genetics 1990 Position paper on genetic discrimination. Reprinted in *Reproductive and Genetic Engineering* 3(3): 287–95.
Duden, B. 1993 Visualizing 'life'. *Science as Culture* 3, pt 4, no. 17: 562–600.
Farrant, W. 1985 Who's for amniocentesis? The politics of prenatal screening, in Homans, H. (ed.) *The Sexual Politics of Reproduction*. Gower: Aldershot.
Green, J.M. 1990 Calming or harming: a critical review of psychological effects of fetal diagnosis on pregnant women. *Galton Institute Occasional Papers*, second series, no. 2.
Green, J.M. 1995 Obstetricians' views on prenatal diagnosis and termination of pregnancy: 1980 vs 1993. *British Journal of Obstetrics and Gynaecology* 102: 228–32.
HFEA (Human Fertilisation and Embryology Authority) 1994 *Donated Ovarian Tissue in Embryo Research and Assisted Conception*. Public Consultation Document, HFEA: London.
Jordanova, L. 1993 Gender and the historiography of science. *British Journal for the History of Science* 26: 469–83.
Lippman, A. 1994 Prenatal genetic testing and screening: constructing needs and reinforcing inequities, in Clarke, A. (ed.) *Genetic Counselling: Practice and principles*. Routledge: London and New York.

Marshall, J.R. 1993 Review of *Schizophrenia: Manifestations, incidence and course in different cultures*. A World Health Organisation Ten-Country Study (A. Jablensky et al., Cambridge University Press, 1992). *Clinical Psychology Forum* 41–2.

Marteau, T.M. 1994 Antenatal testing: a research perspective. *CERES News* 13: 5–6 (Consumers for Ethics in Research: London).

Miller, S.K. 1993 Gene hunters sound warning over gay link. *New Scientist*, 24 July, pp. 4–5.

Mohun, J. 1987 Faulty genes lead to old age. *New Scientist*, 29 October, p. 39.

Nuffield Council on Bioethics 1993 *Genetic Screening: Ethical issues*. Nuffield Council: London.

Radford, T. 1995 Liberal backlash at 'genes fallacy'. *Guardian*, 20 February.

Spallone, P. 1992 *Generation Games: Genetic engineering and the future for our lives*. The Women's Press: London; Temple University Press: Philadelphia.

15 Learning Difficulties and the Guardians of the Gene

Chris Goodey

Genes exist. But how do we know? Because there is a technology that has told us. Scientific techniques led us to genes, and they determine the way we talk about them. Technology seems to drive our ethical and thus our political thinking about them. For example, chorionic villus sampling (CVS) is a single technique which can be used to locate two relatively common conditions: cystic fibrosis (CF) and Down's syndrome (DS). What, apart from the technique itself, do they have in common? Cystic fibrosis interferes with the vital operations of breathing and digesting; it causes what we perceive as physical suffering. These are the grounds on which the justifiability of a termination of pregnancy is open for discussion. Down's syndrome, on the other hand (usually caused by an extra copy of chromosome 21, i.e. trisomy-21), involves the empirical observation that nobody with an extra chromosome 21 in every cell has yet – at least as far as we know – been capable of passing an 'A' level examination. Is this suffering? (The minority of cases where there are added physical complications is a separate matter.) Clearly, there has to be a completely different discussion about the justifiability of terminations of pregnancy in this case, but the differences between CF and DS are smothered by the fact that we associate them with a single technique that identifies both conditions.

This confusion exists at the micro-ethical level too. There are some grounds for saying that the idea of informed choice is adequate for the highly individual nexus of relationships, cultural beliefs, opportunities, etc., surrounding the carrier of a faulty CF gene. Potential parents may be faced with 'difficult decisions' and 'grey areas', but they are grey specifically with respect to physical suffering and caring for physical suffering. Suffering is what we do not like, by definition (monks like wearing hair shirts, but only to reach heaven by liking what they do not like), and suffering is what the choice is about. But the lack of academic ability in DS is clearly not something we dislike in this same fundamental sense; nor can we say that people suffer a lack of understanding or learning ability in the same way that they suffer pain. Perhaps our dislike – where it exists – is a historical and social contingency. If so, then 'informed choice' is really a way of dumping on individual parents the moral responsibility for choices already made by the wider society, and perhaps the burden of challenging those choices.

What, specifically, would constitute the grey area or slippery slope between terminating a Down's syndrome foetus (genetic testing = good) and terminating a female or a black foetus (designer babies = bad)? Not the degree to which you could cope with physical suffering in yourself or your offspring, but a quality – intelligence – which, like medieval notions of honour, depends entirely upon how others perceive it, and which has already been defined for you. The information subsequently given to an individual parent is thus incapable of being neutral or disinterested; and the very belief in grey areas here makes 'informed choice' indistinguishable from the eugenic view that all DS foetuses should ideally be terminated, whereas no normal female foetuses should be. Individual choices are informed by existing (Western) social norms entangled with a stupidly normless technology.

Perhaps we should simply ask people. This is what anthropologists do with most cultural groupings. Can people with Down's syndrome, and others with 'learning difficulties' associated with genetic conditions, be called a cultural grouping? It is difficult to see how they could not, since the social structures actually force them to associate with each other in segregated schools, training institutions and day centres, as well as to organise for their own solidarity in People First and Young People First. Do they suffer? Since we segregate ourselves too, and they are not usually invited to our academic gatherings, it may be difficult, but were we to ask them, I think they would hotly deny that their particular genetic expression of humanity involves any suffering beyond that which is designed for them by other people. Our applications of genetic technology assume we have the prerogative to deny that denial.

Physical difficulties such as CF, sickle cell disease or thalassaemia have some clear trans-historical basis; it would be hard to tell a sufferer there is no essential difference between natural and social science, and therefore that their pain is a mere social construction. But the suffering these conditions involve, beyond any necessary scepticism about cultural perceptions of the word, can only leak from here into the notion of intelligence and learning difficulties from the stand-point of the specific culture that invented the techniques of genetic discovery and therefore the gene. It is the vice-like grip of domination of a single discourse about genetic testing that provides cover for the transfer of 'suffering' from a place where, in general, it belongs to a place where, in general, it does not belong.

Other ways of justifying termination follow the same pattern. For the developers of the technology, what the foetus with CF and the foetus with DS have in common is that their coming to full term, to be born alive, would cost the community more for the duration of their lives than would the statistically normal foetus. In this we hear, not a preconception about the comparative values of pain and learning difficulties, but just the counting of beans. But both approaches are the result of letting values be determined by a reification (an intellec-tual concept being treated as a physical reality in the external world) – the 'gene', created by ourselves in the shape of the blind god/scarecrow of medical technology and held aloft by the global companies which fund research, naturally for a profitable return.

The technology fetish has a totalitarian force which feeds on our passive adoration of it; we receive its blessings on the condition that we stop thinking. So much for bioethical notions of autonomy. The first stirring of a genuine autonomy would be to ask who is in control, to ask questions about the specific culture that surrounds genetic technology. Anthropologists call this 'studying up'. They could ask the guardians of the gene: What are your values? What incantations, spells and rituals belong with your fetish? Who are your significant ancestors?

CULTURE

Before doing this, we need first to make sure that we do not reify 'culture' too. Our everyday usages of this word, and perhaps many of our academic ones, lack a dimension, which can be supplied by a consideration of learning difficulties. The philosopher Ludwig Witt-genstein can help us here. First, he claimed that the mind as we envisage it – a place where thoughts are stored in the form of ideas about the real world before they are given expression in language – is an 'occult medium',[1] rendering vacuous the whole discipline of psy-chology based on it. Rather, there is just language itself, and its meanings are derived simply from its use in particular contexts; its stability, its rules, exist only as a set of various 'language games', which resemble each other in the way members of a family do but not as indicators of some separately existing knowledge that is systematic and certain. Second, he asked what could still be described as essentially human if language, characteristic of our species, could not be said to hook up with some kind of certain knowledge in this way. From this came the famous phrase, 'If a lion could talk, we could not understand him.' It is true that human knowledge and abilities are 'anthropological', variegated and relative; however, this does not mean that they are not unique and essential to humans. Rather, the set of language games is the sign of a specifically human 'form of life', with which the leonine form of life, for example, is mutually incommunicable. Absolute knowledge is not possible; culture matters.

But what matters is something I would term 'deep culture', a deep cultural difference from other forms of life which makes the differences between the various forms of human culture only secondarily problematic.

One of the props to Wittgenstein's argument is his discussion of a group he calls the 'mentally deficient' or 'feeble-minded'. He may have been led to think about them by Ellen Pinsent, who was a prominent advocate of segregating children with learning difficulties into separate schools, on the (eugenic) grounds that they should not be encouraged to form relationships with other children. She was the mother of Wittgenstein's close friend David Pinsent, and their kinship group includes several famous physicians, including Dr John Langdon Down. On this point, however, Wittgenstein turned his back on his own tribe. They were determined to establish biological difference and inferiority. For him, the utterances of this putative sub-species, the 'mentally deficient', however incomplete, ineffective or difficult they may be, are recognisably part of the set of human language games, and therefore of a specifically human form of life. Thus it is the very existence of such people that positively justifies the unity of the human species and the possibility of a mutual understanding, prevailing over anthropological relativism but not asserting itself as a canon of absolute knowledge. (We may add: neither a Western canon in relation to non-Western cultures nor a canon of 'intelligence' in relation to people with learning difficulties.)

This deep culture has species boundaries established precisely by the inclusion of people with learning difficulties. In developing this idea, I do not want to dismiss the importance of the 'relative cultures' lived

by millions and studied by anthropologists. Just the opposite: I think
deep culture can enhance them. The problem for relative cultures is:
how can we find room for the discussion of them without the discussion itself being the very means by which the politically dominant tribe
and its culture are left undiscussed? To talk about 'ethnicity' is automatically to talk about a particular form of power and oppression
without being able to extricate ourselves from its grasp; the term
assumes there is still 'us and them', or 'us and cultures'. But the
alternative – to look at the dominant culture as just one among
many – seems to lead to a loss of all our bearings in an amorphous
and competitive relativism, ignoring the realities of power. We can
draw a parallel with familiar political debates about racism, sexism,
heterosexism, etc., and the competitive claims they may have on each
other (such as the homophobia or sexism of some men otherwise
discriminated against themselves, or indeed some physically disabled
people's distancing of themselves from those with learning difficulties[2]). At the same time, there is a general notion of justice which
runs through all of them and so can include all without reducing
justice to the notion of equality, which, in a society with structural
racism, sexism, heterosexism, etc., would simply duplicate the inequalities of existing power structures. In the same way, the notion of a
unified and specifically human or 'deep culture' allows us to speak
about what we have in common as humans without liquidating or
anathematising the differences among us.

'DIFFICULTY' AS A CULTURAL NOTION

This attempt to demystify 'genes' and 'culture' allows us to look at the
curious cultural standpoint which is none the less curious because we
ourselves are it: the standpoint which regards people with learning
difficulties as excluded from the bounds of mutual recognition and
understanding, and in this sense non-human. This culture exists at two
levels. First, it is the three centuries of Western-civilisation-as-we-
know-it, including the present with its prenatal eugenics, as well as
the postnatal variant of the late 1930s and early 1940s. Second, within
this, there is a very restricted tribal group, a subculture (with strong
genealogical ties) of medical professionals.
    However dominant this standpoint appears, I have also encountered
widespread ignorance of it. Palaeopathological evidence shows that

prehistoric nomadic groups in places such as Australia and Lower California included adults with congenital multiple physical and mental disabilities.[3]

Interviewing parents with 21-trisomic children in East London,[4] I came across more than one who was surprised to hear from me, for the first time, that Down's syndrome was labelled by the outside world a *mental* handicap; not having themselves grown up with the idea that going to Oxford or Cambridge or medical school is something desirable or even makes any sense, it had simply never crossed their minds that there was something 'wrong' with their child in that sense.

What, then, are the characteristics of the specific culture that sees its human 'problems' or 'difficulties' as something to do with the capability for academic learning? For most people and at most stages in history, the word 'difficulty' is simply synonymous with 'being alive'. The problems begin when a position of class dominance leads to a separating out of 'difficulties' from actual existence. Just as utopias are the effluence of a society with an excess technical capability (for example, genetic technology), so we create a dystopia too, a fantasy world in which we conceptualise certain difficulties with the aim of disposing of the people who embody them. Alongside the real world, we create an imaginary wrong world, and populate it with wrong kinds of people who must be got rid of. Say I am a consultant obstetrician or paediatrician; my cleaning lady returns to Manila in the middle of the afternoon without giving notice, or my wife decides to resume a full-time career after 15 years of voluntary work, mainly in my own home. What on earth am I to do? This is already a difficulty. How much more of a difficulty must it be to have a child who will not go to Oxbridge or follow the tribal route into my old medical college. Such a child is an anomaly because it interrupts a genealogical line. My membership of a wider dominant group imposes on me the duty to spread these values – actually an exotic subcultural inhibition or taboo – as far as possible through the rest of society, for its own good. The wider group's overall interest is the worship of wealth and the maintenance of its own power, for which 'intelligence' has a talismanic importance. The professional subculture, though fairly small, has the job of secreting this value about the acceptability of offspring, like a rare orchid with a particularly powerful fragrance.

This is achieved by means of rituals, which other elite groups have also used. In classical Athens, a week after the birth of a child to a well-to-do family, and after the father had consulted with the midwife about the potential health and development of the child, he would run round the central hearth five times in a postnatal ritual called *amphidromia*, symbolising the acceptance of the infant by the household. We do not know how these consultations went, or what the link was between them and the ritual acceptance or non-acceptance. Nor do we know exactly what is the connection between what our genetic counsellors say and the prenatal ritual of CVS, where the technology has, in a sense, already determined the possible outcomes.

There are some direct historical links with our own society that can be made. Christian baptism is a ritual to exorcise the devil in a newborn child, and when Luther recommended infanticide for older children whom we would call autistic, he did so because the ritual had failed. When John Locke spoke about the desirability of infanticide for idiot infants on the basis of his new psychology, he constructed these idiots on an analogous theological basis. In fact, psychology's concept of what was first called idiocy, then mental deficiency, then mental handicap and now learning disability, as a general lack of intelligence, is no more than three and a half centuries old. It was invented by politicised theologians as part of an emerging ruling-class ideology of political autonomy and consent; they secularised the Calvinist doctrine of the 'elect' into a bourgeois theory of active intellect, replacing the passive 'illumination' of baptism.[5] In a mass society, the 'elect' have become everybody – or nearly everybody.

What we consider to be our autonomy, our rights, our freedom to consent, depend for their existence on the exclusion of those deemed to be incapable of autonomy, rights and consent; this is not a social exclusion but implicitly still an exclusion from the species – hence the drive to associate such incapability with physiological differentiation. Only this kind of marked difference, highlighted by the mind–body dichotomy (itself a construction made in that same 350-year time span), can justify our attitude to those thereby excluded. Learning difficulties, then, are not trans-historical, and they differ from physical disabilities in this respect.

My phrase 'the wrong world' is borrowed from the title of a painting by Breughel. One of its details is a woman with a hooked nose and chin tying up a baby as tightly as she can, glaring at the infant malevolently. The infant looks back at her – with exactly the same physiognomy and expression. There is an obviously disturbed relationship here.[6] Breughel's wrong world is a dystopian vision of all the difficulties imaginable in relationships; but he seems to ask us: Is this the best way to deal with difficulties? Do we not rather, in this way, create difficulties for ourselves, partly as a result of the leisure we have to imagine them? Our imagined difficulty is what creates and embodies that difficulty in the child. A society and a class whose self-esteem derives from the opportunities afforded by global economic and cultural exploitation have leisure for the dystopian vision, but also a *need* for dystopia, the imaginary picture of something 'wrong' with other humans. This culture of dystopia may be said to be a pathological phenomenon, and in the broad historical and sociological perspective, it is an extremely unusual one. When Professor Wald says, as he did in *The Times* during a discussion of the problems caused to parents by the (considerable) degree of inaccuracy in CVS, that prenatal 'anxieties may be the inevitable by-product of benefits to society', he is presupposing that everyone will agree with him that the termination of a Down's syndrome foetus is a benefit, an assumption that has anthropological origins in his own small subculture.

The alleged benefits are to the whole society, not to the prospective unterminated foetus. The professor's 'researchers have costed their screening method and estimate that it costs £38 000 to avoid one baby with Down's syndrome. The lifetime costs of caring for one child with Down's syndrome have been estimated at £120 000', says the *BMJ*. On a broader utilitarian view, 'the most important reason for screening...is the avoidance of...distress to the families concerned'. I have already suggested that the added costs – presumably those of adoption and segregation, otherwise the figures are just meaningless – are simply the enactment of certain subcultural values. Professors, researchers and people like ourselves have a circular connection with rational choice theories; when we are sold defective goods we have the confidence to send them back. The lifetime costs (either in cash or in anxiety and distress) are estimated on the basis of what we would do, and of what we, as professionals and academics, teach others to do and feel.

Marginal ethical decision-making is the realm of politics. Could we use each £38 000 more effectively somewhere else in the NHS? The short answer must be yes, but I am more concerned with the broader arguments about alleviating anxiety and distress, and these, too, are political. It ought to be obvious that to challenge existing criteria about the appropriateness of termination for a foetus with potential learning difficulties (that is, a kind of difficulty our culture disapproves of), is not the same thing as to argue that it has an inviolable sanctity. I want simply to concentrate on utilitarian issues of the quality of life and the satisfaction of preferences.

Any bioethicist who perceives these congenital difficulties as necessarily indicating a low quality of life for subjects, carers and society in general is making an unjustified assumption. And regarding termination as a benefit is perhaps part of an attitude that helps to create socio-political difficulties.

LEARNING DIFFICULTIES, RACISM AND POLITICS

For an everyday political angle on genetic testing and screening that might start to reflect this view, I have looked in vain in the pages of the 'serious' press. In the end, I found something in the *Sunday Mirror*. A 1994 editorial about designer babies, jostling for space with scandalous vicars and bribed football managers, said:

It is already possible for parents to choose the sex of their children. It will not be long before a 'designer baby' can be created to order. Picture the scene at the clinic as parents choose the perfect child. Eye colour? Blue. Skin colour? White. IQ?[7] High. It is a chilling prospect that should stir a deep, instinctive unease in us all.

Is this just tabloid rhetoric? Perhaps prenatal eugenics and the postnatal Nazi racial and euthanasia policies evoked are not morally equivalent;[8] perhaps we should be using reason as well as instinct. What is notable, however, is that the editor fails to see any moral difference between choosing skin colour and choosing intelligence. Me too. On this issue I am with the prehistoric nomads of Lower California, the working-class families of East London, the *Sunday Mirror* and Ludwig Wittgenstein; I am against Western civilisation as we know it, the ethics committee of the BMA and the broadsheet journalists.

The real difficulties and the real pathology are, as I have said, those of the culture which labels the 'learning difficulties'. This is the same culture which, for the same reasons, both sustains itself and creates problems for itself by provoking racism and subordinating 'relative' cultures. 'Ablism' (at least where learning difficulties are concerned) and racism are historically the same thing. The 17th century was rife with demonological explanations for the appearance of 'idiot' children and the non-white peoples of the New World. The psychology of idiocy in the late 18th century was raided by Dutch physicians to establish a scientific racism based on a hierarchical scale which described non-whites for the first time as biologically inferior in intelligence to whites.[9] The same process in reverse was invented a few decades later, when medical anthropologists explained 'mongolism' as the recapitulation and arrest in the womb of an earlier and inferior stage of human racial development. Recognition of this should be at the forefront of our current concerns. We need only look at the resilience of fascism, at the organised attacks by German neo-Nazis on disabled people as well as non-Europeans, and even more pointedly at the resistance itself: contingents from People First have taken part in all the major anti-BNP (British Nationalist Party) demonstrations in Britain.

Politics involves social problem-solving, which means knowing where the problem lies and whose problem it is. Underpinning this skill there are intrinsically political decisions to be made about the cultural questions 'What is it to be black?' and 'What is it to be a woman?' in this society. But alongside these there is the issue of deep culture: 'What is it to be human?' The social construction and prenatal destruction of the creature with learning difficulties are cultural issues like the first two, but they also provide an underlying context and natural-history boundaries within which the first two can be challenged. Help subvert the difficulty of learning 'difficulties' and you might get somewhere with the others. Conversely, if elimination is all right for learning difficulties, it can always be all right at some stage for being black or female.

Members of the ethics committee of the BMA think that the question 'What is it to be human?' should be determined not, of course, by prehistoric nomads, eminent philosophers or working-class East Londoners, but by the Chief Secretary to the Treasury on behalf of the IMF and the World Bank, who follow criteria bequeathed to them by their ancestors, the Poor Law guardians, in answering the question: What are the fiscal consequences of preserving this detritus?

Our argument against this is practical. Creating difficulties creates difficulties: it destroys social cohesion and integration, just as structural racism and sexism do. Thus, the problem with the sub-utilitarian approach is that bean-counting is simply impractical, utopian and counter-productive. This applies equally to the good intentions about informed choice or avoiding distress; it is not only the unacceptable but the acceptable face of genetic counselling that takes as read a hierarchical scale of values attached to the fictions of intelligence. This scale is just the predisposition of a particular culture. If anxiety exists about there being a grey area between destroying an XX foetus just because it is XX and destroying a Down's syndrome foetus just because it has an additional chromosome 21, what is it that causes the anxiety in the first place, and is the anxiety necessary? If, rather than assume that a positive-looking CVS means amniocentesis and a termination, we want simply to make sure that parents are fully informed, this begs an enormous question. What is it that informs the information we give them? Is the information neutral? Or is the information itself the story told by a particular tribe with its technology fetish? Until this question is opened up, I see no difference between prescribed eugenics and informed choice.

If policing the underclass is important, we shall find a 'criminal' gene. If policing love and friendship is important, we shall find a 'gay' gene. If allowing rational choice and privatisation to flourish is important, we shall not even look for a 'selfishness' gene (on the contrary, every gene will, by definition, be selfish). Our counselling is driven accordingly: finding is preceded by seeking, and seeking is preceded by motivation. The first consultation by genetic counsellors therefore needs to be with themselves. What are my values? What is intelligence to me? What do I esteem about myself? The second consultation follows from realising that this is one 'I' among many who need to consult with each other, and who are by no means confined to the expert technicians of a monolithic subculture. To adapt a famous quotation from Plato about philosophers and kings, it is not until genetic counsellors become politicians and politicians become genetic counsellors that we shall be anywhere near beginning to find a solution to the problems we find in the real world. The discussion belongs in the public arena. It should not remain hidden in the bowels of the corporate state and the corporate NHS, protected by the top-secret label 'science and technology'. Nor should anthropologists and moral philosophers remain in their armchairs calling for the gene genie to be put back in the bottle; technology, as such, does not submit to being banned. It is the essentially political questions primordial to the technology that need discussion; and in making decisions about who are acceptable to us as fellow citizens, we need to consider ourselves as citizens too, not just as biochemical and ethical experts.

A fundamental public discussion of the values attached to intelligence and learning difficulties not only needs a public arena but would threaten to create and enhance one. A space might be created in which the endlessly resuscitated agendas of race-and-intelligence, class-and-intelligence, etc., could be sidestepped, and in which other than selfish values could emerge. There is of course always the possibility that 'taxpayers' careful with their money would have their say, having been persuaded that the application of genetic technology in this area would lead to a financial saving.[10] But those of us taxpayers with an eye on the necessary realities of cooperation and mutual dependence will have our opportunity to say something different. In discussing learning difficulties together, we learn something about the basic questions of culture: What do we want to be, and how will we belong?

NOTES

1. The relevant passages are in the first volume of Wittgenstein's *Remarks on the Philosophy of Psychology* (Blackwell: Oxford, 1980).
2. See, for example, the remark of Simone Aspis, a member of People First, that physically disabled campaigners themselves see people with learning difficulties under the medical model, in *DAIL Magazine: Disability Arts in London* 98 (1995).
3. Relevant material may be found in the subject index of *Paleopathology Newsletter* (1983).
4. *Living in the Real World: Families speak about Down's syndrome* (The Twenty-One Press: London, 1991).
5. See my 'The psychopolitics of learning and disability in the seventeenth century', in Anne Digby and David Wright (eds), *Historical Perspectives on People with Learning Disabilities* (Routledge: London, 1995).
6. I am grateful for this observation to Inge Mans, who has published in the Netherlands on the early social history of people with learning difficulties.
7. IQ is, of course, already being selected by amniocentesis – a fact overlooked by others, too, whom one would expect to be capable of thinking harder.
8. Michael Burleigh's book *Death and Deliverance: 'Euthanasia' in Germany 1900–1945* (Cambridge University Press: Cambridge, 1994) gives a detailed account of what happened to people with various disabilities during and preceding the war.
9. Norris Saakwa Mante's work in progress, 'The emergence of race in Britain, 1770–1850' (Harvard Ph.D. dissertation), deals with this aspect of the history.
10. There are certainly scientists around who are prepared to trade in these iatrogenic anxieties. See, for example, J. Maddox, 'The case for the human genome', in *Nature* 352 (1991). This example was pointed out to me by Audrey Tyler.

16 Matchmaking in the Clinic: Gamete Donation and the Management of Difference

Frances Price

The novel techniques of genetic medicine have stimulated both scholarly and popular interest in 'genetic information', and fuelled controversies about 'communication', 'privacy' and 'consent'. In recent years, genetic tests of embryos, foetuses and prospective parents have introduced molecular techniques into the clinic. This has generated widespread public debate and substantial research funding, and has heightened popular expectations about genetic self-knowledge. Yet in one expanding area of reproductive medicine – the egg and sperm donation programmes now flourishing in Britain and elsewhere – issues surrounding genetic information have so far received scant attention (Price 1995).

Gamete donation is often presented as comparable to organ or tissue donation, yet human eggs and sperm, unlike kidneys and blood, establish a genetic link with a new generation. Who should know what when genes are in transition between persons in a culture of concealment – given that anonymity and secrecy surround the recourse to donor gametes? What information should be collected and communicated about gamete donors and recipients?

The issues opened up by the contemporary use of donated human sperm and eggs to facilitate pregnancy are likely to become more pressing as knowledge of the human genome increases. Men and women are now being actively recruited as sperm and egg donors, to provide their gametes for other people's procreative projects (Cook & Golombok 1995). Widespread publicity given to this particular treatment service in recent years has led more women and men to consider that access to a donation programme could provide a solution for them in their quest for a child. At the same time, there is a perceived need for ever more stringent screening of donors for genetic disorders or infectious diseases, but no consensus on what information to gather and what to share with the parties to a donation or with any child born of the donation. What might be regarded as an 'acceptable risk' in relation to in vitro fertilisation when only the couples' own gametes are used may be perceived as problematic when it involves not only a doctor but also a donor.

As clinicians recruit and classify donors and strive to 'match' them to one or more recipients, they assess the genetic make-up of donors and recipients and categorise them to some extent in various ways. They also make judgements both about what information to collect to effect a 'good-enough' match and about what to convey to the participants in the donation. Thus, the medical context and legislative framework in which gamete donation now takes place shapes practice in ways that need to be explored. In this paper, I focus on deliberations about 'genetic risk' and 'ethnic group', and explore the problematic nature of these assessments and communications about them.

BACKGROUND TO DONATION PROGRAMMES

Demand for donated sperm and eggs to facilitate pregnancy has increased rapidly in recent years, despite the taboo surrounding sperm donation in Britain. Even in the 1930s, as male infertility began to be acknowledged as a factor in reproductive failure, it was not public knowledge that a small group of doctors were offering donor insemination (DI) to their patients. In the face of moral, theological and legal opposition, the practice was conducted in secret – so much so that the first published account of British clinical experience with DI in the *British Medical Journal* in 1945 created a furore (Barton et al. 1945, Pfeffer 1993). Three years later, a commission of inquiry appointed by the Archbishop of Canterbury urged that DI be made a criminal offence. In 1960, a second commission, the Feversham Committee, stopped short of advocating the criminalisation of DI only for fear of driving the 'unwholesome' practice underground, and recommended that it should be provided only discreetly to discourage its spread (Feversham Committee 1960). Perceiving DI as a threat to the stability of the peerage and the monarchy itself, Feversham concluded that: 'Succession through blood descent is an important element of family life and as such it is the basis of our society.' The report added: 'Knowledge that there is uncertainty about the fatherhood of some is a potential threat to the security of all.' Despite this, the demand for DI increased, bolstered by the relative absence of treatments for male infertility. DI programmes only became available within the National Health Service, however, following the Peel Report in 1973 (Peel 1973).

The use of donated eggs is much more recent, and became possible with the techniques that were developed as part of in vitro fertilisation (IVF). An Australian IVF team reported the first pregnancy from a donated egg in 1983 (Trounson et al. 1983). In 1984, the Warnock Committee of Inquiry into Human Fertilisation and Embryology accepted egg donation as the logical parallel of the now well-established sperm donation programmes (Warnock 1984). In line with sperm donation, the committee recommended that egg donors should be anonymous and have no subsequent rights or responsibilities with respect to the child. The Warnock Report formed the draft framework for the Bill which became the Human Fertilisation and Embryology Act of 1990. This provided for a new statutory licensing authority, the Human Fertilisation and Embryology Authority (HFEA), which now licenses embryo research, the storage of gametes and embryos, and all those types of assisted conception which involve in vitro fertilisation and those which involve donated gametes or embryos (Morgan & Lee 1991).

Donated eggs were initially intended for women with premature ovarian failure or who were born without ovaries. Yet the 'indications' for using donated eggs soon multiplied: today, they are also given to women who ovulate but have failed to conceive through IVF, to older women who have passed the menopause and, in a few instances, to those who seek to avoid passing on a genetic disorder such as Huntington's disease (Davies 1992). Recruitment campaigns have become more extensive – ranging from posters in GP surgeries to appeals in local newspapers, women's magazines and on breakfast television – yet the number of women who donate eggs continues to fall far short of satisfying this growing demand (Horne et al. 1993).

Egg donation differs from sperm donation not least in its attendant medical risks (Cameron et al. 1989, Smith & Cooke 1991, Davies 1992). Prospective egg donors must consent to regular hormonal injections to stimulate the development of many eggs in a single cycle, and then undergo a surgical procedure, usually under general anaesthesia, to remove mature eggs. Financial incentives for sperm and egg donors, now the subject of renewed controversy, also differ (Vines 1994). Sperm donors, perceived as donating primarily for financial reasons, are offered in the region of £15 plus 'reasonable expenses' for each donation, whereas women are generally offered reimbursement of their travel expenses or 'in kind' benefits, such as free sterilisation or IVF treatment. Recently, however, some clinicians have argued that women should be offered financial compensation for the time and effort required to donate their eggs.

A furore about 'eggs for sale' was provoked by a BBC 1 *Here and Now* programme on 'Human Hens', broadcast in October 1995. The programme reported that egg donors were being recruited on behalf of at least one HFEA-licensed London clinic by Hope, a private agency based in Cornwall, which links donors with recipients for a fee. 'White' donors receive £850 while 'rarer' 'black, Asian or Chinese' donors are paid in the region of £1000. Newspaper accounts claimed the case exposed a 'loophole' in the legislation, which bans payments by clinics but not by agencies such as Hope (Fletcher 1995, Harwood & Lakeman 1995).

ANONYMITY AND THE ORIGINS OF 'MATCHING'

Part of the explanation for the rapidity and ease with which gamete donation has become an accepted part of contemporary reproductive medicine lies with the culture of concealment that has grown up around the practice. Secrecy and anonymity became the linchpin of medical practice using donated gametes in the days when only artificial insemination was technically possible. Perceiving male infertility to be particularly stigmatised, medical bodies such as the Royal College of Obstetricians and Gynaecologists advised that neither the donor nor the recipient couple should know the other's identity (RCOG 1983, Golombok & Cook 1993). Anonymity would also prevent the intrusion of a 'third party' into the couple's sphere and would support the stability of the 'family'.

As donor insemination became widely available as a service in reproductive medicine, the practice of 'matching' a sperm donor to a recipient was presented as a crucial part of the culture of concealment. The RCOG, for instance, recommended that donors should be carefully selected. The intention was 'to ensure intelligent, fit, healthy donors who have no family history of genetic disease'. The Royal College also advised that centres should try to use donors with the same physical characteristics as the husband, and to match the sperm donor and husband on three stated criteria: eye colour, hair colour and blood type. In practice, height and body build are sometimes also taken into account. These visual characteristics were apparently chosen not only to enable the child to 'pass' as the genetic offspring of its social father, without provoking comments from outsiders, but also on the basis of popular knowledge of genetics: children typically discuss the inheritance of eye colour, hair colour and blood type in the classroom.

Interestingly, the ruling that donations should be anonymous has been relaxed in gamete transfer between women (Haimes 1993). Would-be egg recipients are increasingly making their own arrangements, recruiting sisters or friends to act as personal egg donors (Price & Cook 1995). Some clinics maintain anonymity but let a woman waiting for egg donation move up the queue if she recruits an egg donor for another woman.

GENETIC SCREENING OF DONORS AND RECIPIENTS

Contemporary best practice in gamete donation programmes entails screening the donor for certain transmittable infectious diseases, but the issue of the most effective method of screening is very much open to debate (Barratt & Cooke 1990, Barratt 1993, American Fertility Society 1993). Clinicians in the field practise some genetic selection, although Barratt points out that donor screening for genetic diseases has in the past been 'grossly inadequate' (Barratt 1993: 9). Assessments of genetic make-up are largely based on information provided by prospective donors about their family history of diseases such as asthma, heart disease and diabetes, and there are few guidelines concerning the procedures used to elicit a detailed history. Barratt recommends that:

> the screening should be performed by a qualified medical person who is aware of the minimal genetic screening for donors. It is important to remember that the level of competence at questioning is extremely important in obtaining the genetic screening by history, e.g., Timmons and colleagues, using extensive data, interview and collection systems, found that 5.4% of 149 donors had self-assessed genetic disorders: elsewhere extensive investigation revealed that 29.5% had conditions within the family which warranted further investigation, a 5.5 fold discrepancy. (Barratt 1993: 9)

Direct genetic tests are rarely carried out; few clinics test donors for cystic fibrosis carrier status, for instance. Cytogenetic tests to detect chromosomal abnormalities may sometimes be carried out. Guidelines produced by the British Andrology Society (BAS) currently state that karyotyping of donors 'would appear advisable', but this is 'not considered mandatory' (BAS 1993). The society advises: 'Whether or not to karyotype a donor depends on one's judgement of how small a risk one can afford to ignore.'

Opinions also differ as to whether screening should be extended to the recipient – to allow matching of donor and recipient in an attempt to avoid the accumulation of risk factors for either genetic-based or acquired diseases in a child born of the donation. For instance, some clinics match sperm donors positive for cytomegalovirus to recipients who also test positive for the virus (Barratt 1993).

The HFEA's Code of Practice requires that centres 'should as minimum' follow the BAS's guidelines for sperm donors (HFEA 1993). No official body has yet produced guidelines for the screening of egg donors, but the HFEA states that 'centres should adopt whatever is current best practice in the scientific testing of semen samples and of donors of gametes'. The Code of Practice sets only minimum restrictions on who can receive or donate eggs, taking into consideration 'the welfare of any child who may be born as a result of the treatment (including the need of that child for a father), and of any other child who may be affected by the birth'. However, the Code does stipulate that egg donors must be at least 18, and that they must be under 35 unless there are 'exceptional reasons'. It also states that centres should, when deciding whether to accept a donor, consider any personal or family history of heritable disorders, any personal history of transmissible infection, whether the donor has children of her own, and 'the attitude of the donor towards the donation', although no guidance is given as to how any of these considerations should be put into practice.

As required by Section 31 of the 1990 Act, the HFEA also collects and stores in a central register identifying information about egg (and sperm) donors, recording their age, date of birth, height, weight, 'ethnic group', eye, hair and skin colour, and whether they have children of their own. It also asks donors to give their occupation and 'interests'. In an optional final section of the 'donor information form', each donor is offered the opportunity to provide: 'a brief description of yourself as a person. This should be something which could be passed on to any child born as a result of your help. It may also be seen by the parents.' A number of donors, however, choose not to complete this section.

The HFEA provides donors with a rationale for the collection of this non-identifying information in its leaflet 'Sperm and Egg Donors and the Law'. The authority states that this information:

> may be helpful to children born following treatment with your eggs or sperm who want to gain some insight into their genetic origins. With this in mind, the Authority collects general information about what you look like. It also offers you the opportunity to describe yourself and your talents and your interests more fully in your own words if you think that could be helpful to a child born as a result of your donation. If Parliament in the future makes this decision, such a child after reaching the age of 18, could contact the HFEA for the information.

Thus, although donors are informed that, by law, they will not be the legal parent of any resulting child, and that they have relinquished all legal rights and claims over offspring that may result from their donated eggs, they are asked to consider the psychological and genetic wellbeing of the child. Furthermore, donors are warned of 'the possibility that a child born disabled as a result of a donor's failure to disclose defects, about which she knew or ought reasonably to have known, may be able to sue the donor for damages'. Donors can be pursued under the Congenital Disabilities (Civil Liability) Act 1976 if they knowingly conceal a family history of genetic disorder.

In this regulatory climate, gamete donors face a variety of risks and uncertainties from the genetic, medical and personal information provided by them, and tacit social judgements based on this information. Communications about disease or genetic make-up can fundamentally impinge on perceptions of identity and lifestyle and on existing social commitments. In a recent study of egg donors (Price & Cook 1995), one donor, a single, childless, 20-year-old medical student, noted:

> I was slightly worried about the wording on some of the leaflets saying I could be sued if I did not reveal any inheritable medical conditions that I ought 'reasonably to know about'. I appreciate this is to protect any child born after your eggs were used, but it doesn't specify how much you need to know and which relatives to ask. I have asked both parents about their sides of the family, but feel that more guidance should be given. For example, do I need detailed

medical histories from only close relatives, or everyone in my whole family (which is completely unfeasible). On contacting the HFEA, they just sent a letter saying I ought to mention to [the clinic] all diseases that I know of, which of course I did, but I still feel a bit unhappy about the lack of help/guidelines, and at one point considered not being a donor.

Another egg donor, a 35-year-old solicitor with two children, listed as one of her several concerns about egg donation a growing preoccupation with genetic issues:

It seems possible that in the future it might be decided that the child has a right to know who its 'mother' is, particularly in view of the increasing preoccupation with genes and genetically-transmitted disease, behaviour, characteristics etc.

The prospect, and public anticipation, of genetic screening using DNA tests raises further questions concerning which genetic diseases should be screened for, and by what means. Should the donor be informed of the results? Should prospective recipients also be screened, so that risk factors can be assessed? Several leading clinicians in the field have put forward different views and practices concerning these matters.

Eliciting and assessing family history can be difficult: a proportion of donors do not tell any member of their family of their intention to donate and, like the medical student quoted above, may be unaware of the nature and extent of genetically related conditions in their extended family. Moreover, in the United Kingdom, most of the men and some of the women who act as gamete donors have not yet had children. At the same time, recipients may expect that rigorous screening and 'matching' procedures ensure a 'healthy' child. The uncertainties surrounding estimations of genetic risks in this context are rarely openly acknowledged.

'ETHNIC' DONORS

In general, the idea of avoiding overt 'difference' is deep-rooted in the management of gamete donation programmes in the UK. Anonymous donors assume that they will be 'matched' with recipients (or in certain circumstances, the recipients' partners). Effectively, when a 'good-enough' match is secured, the donor vanishes from view: substitution, to all intents and purposes, has taken place. Moreover, it is pointedly those instances when a 'difference' between donor and recipient (or her partner) is identified as marked that spark controversy. Politically salient notions of 'difference' may be brought to the fore, particularly when donation appears to violate 'racialised' boundaries.

In clinical practice, the perceived need to match donor and recipient can create special difficulties. Simone Novaes has described two contentious cases which came up for review within the centrally organised French sperm bank, Centre d'Études et de Conservation de Sperme (CECOS) (Novaes 1993). The sperm bank was unable to provide donor sperm that satisfied the CECOS criteria for matching blood type in one case, and skin colour in the other. In each case, the recipient couple had requested that CECOS disregard the 'mismatch' and allow the woman concerned to undergo donor insemination.

CECOS ruled in the first case, where the donor blood type was at issue, that the recipient couple could proceed with donor insemination on the grounds that the blood type mismatch was not visible – 'not immediately apparent'. Moreover, the couple concerned had already told the sperm bank of their intention to tell other people, friends and family, that they had sought donor insemination. In the second case, the issue was more than just skin colour. The recipient couple, of Vietnamese origin, faced with a prospect of no CECOS donors of 'appropriate' ethnicity, had requested insemination with 'European' sperm, and had accepted the prospect of a child with features different from their own. Novaes reports that only after a 'lively discussion' was the Vietnamese couple accepted for donor insemination, and then only on a technicality – on the grounds that the CECOS sperm bank could not observe its own 'matching' rules because it did not have sperm from donors from a wide range of ethnic groups. The bank also ruled, however, that its decision in this second case should not create a precedent.

In the UK, medical practitioners and those who license the treatment centres in which they work currently classify their patients and donors by 'ethnic group'. Thus, the HFEA's donor information form provides a choice of nine categories: White, Indian, Black Caribbean, Pakistani, Black African, Bangladeshi, Black other – describe..., Chinese, Any other – describe.... Such classifications are not only theoretically contentious but socially problematic. Thus, in 1994 the *Guardian* newspaper carried a report that a fertility clinic in Nottingham 'has turned down an Indian couple's request for the wife to be implanted with a white woman's eggs. No Indian eggs were available' (Anon. 1994).

Such categorisations within the clinic may come to reaffirm discredited notions of 'race' (Tizard & Phoenix 1993, Macbeth 1993; also see Sharma, Chapter 4). In 1994 for instance, newspapers heralded the claim that a 'British black woman' – identified in one paper as Afro-Caribbean – was about to be given 'a white woman's egg'. This decision was justified, by one of the clinicians at the licensed centre involved, on the grounds that the recipient's partner was said to be 'of mixed race', and thus any child born of his sperm would also be 'of mixed race', whether a 'black' or 'white' egg was used.

The shortage of 'black donors' was highlighted at the time by the *Independent*, who quoted the clinician involved in the case as saying: 'The mother has been waiting more than four years and is approaching her mid-40s. She does not have any time to waste. We did hope for an arrangement with a black donor, but that fell through' (Timmins 1994). In *The Times* story, he was again quoted: 'We are desperately short of donor eggs of all colours and races, but particularly from ethnic [minority] communities' (Gerard 1994; the brackets are in the original).

For some years, IVF practitioners have drawn attention to the lack of 'ethnic minority' egg and sperm donors and have advertised for such donors, using the media to help to promote the recruitment drive. For example, in January 1994 the BBC Radio 4 *Today* team included as news in its programme the pressing need for more sperm and egg donors to help infertile couples to have children. The news item highlighted the shortage of donors from 'black' and 'Asian' people, and attributed this apparent unwillingness to gamete-donate to some distinctive concern with questions of 'ancestry' or family relationships – to 'cultural' difference.

There is something fundamental at stake here, in the concern that ideas about relationships, gender and generation should stand in the way of medical intervention to 'assist' the infertile. Human gametes are coming to be seen as 'spare parts', affording a remedy through 'donation' and substitution, comparable to blood or kidneys. The shortage of particular donors comes then to be seen as a problem of cultural resistance to contemporary medical interventions in reproduction, rather than as a complex, relational issue. 'There is a particular onus on those who speak from the vantage point of the majority culture to make their own cultural foundations explicit', argues

Marilyn Strathern (Chapter 2 in this volume). As Ruth Frankenberg has pointed out, cultural practices associated with 'whiteness' are usually 'unmarked and unnamed' (Frankenberg 1993).

CONCLUSION

In 'unassisted' reproduction, women and men who envisage becoming parents are encouraged to do all they can to ensure that their future child is 'fit and healthy', and this may include undergoing prenatal genetic tests. But in gamete donation, in crucial respects, the focus shifts as the donor and doctor participate in the attempt to establish a pregnancy. Both add new dimensions of risk and uncertainty, and involve new allocations of responsibility and accountability for the consequences of the donation. Few of these issues are prominent in contemporary discussions surrounding either the recruitment of donors or the promotion of this particular 'treatment service' in the UK. The medical discourse of 'matching' – of a medically managed substitution which seems to offer sameness, not difference – has, in a context of secrecy and anonymity, effectively obscured many of the complexities surrounding gamete donation.

REFERENCES

American Fertility Society 1993 Guidelines for gamete donation. *Fertility and Sterility* 59 (supplement 1).
Anon. 1994 Egg request refused. *Guardian*, 12 January.
Barratt, C.L.R. 1993 Donor recruitment, selection and screening, in Barratt, C.L.R., & Cooke, I.D. (eds) *Donor Insemination*. Cambridge University Press: Cambridge.
Barratt, C.L.R., & Cooke, I. 1990 Risks of donor insemination: a review. *International Journal of Risk and Safety Medicine* 1: 113–15.
Barton, M., Walker, K., & Wiesner, B.P. 1945 Artificial insemination. *British Medical Journal*, I: 40–2.
British Andrology Society 1993 BAS guidelines for the screening of semen donors for donor insemination. *Human Reproduction* 8: 1521–3.
Cameron, I.T., Rogers, P.A.W., Caro, C., Haman, J., Healy, D.L., & Leeton, J.F. 1989 Oocyte donation: a review. *British Journal of Obstetrics and Gynaecology* 96: 893–9.
Cook, R., & Golombok, S. 1995 A survey of semen donation. Phase II: the view of the donors. *Human Reproduction* 10: 951–9.

Davies, M.C. 1992 Oocyte donation, in Brinsden, P.R., & Rainbury, P.A. (eds) *A Textbook of In Vitro Fertilisation and Assisted Reproduction*. Parthenon Publishing: Carnforth, pp. 385–98.

Feversham Committee 1960 *Report of the Departmental Committee on Human Artificial Insemination*, Cmnd 1105. HMSO: London.

Fletcher, D. 1995 Would-be mothers pay £1,000 for donor eggs. *Daily Telegraph*, 1 November, p. 13.

Frankenberg, R. 1993 *White Women, Race Matters: The social construction of whiteness*. Routledge: London.

Gerard, L. 1994 Black to receive 'designer embryo'. *The Times*, 25 January.

Golombok, S., & Cook, R. 1993 A survey of semen donation. Phase I: the view of the centres. *Report to the Human Fertilisation and Embryology Authority*. HFEA: London.

Haimes, E. 1993 Issues of gender in gamete donation. *Social Science and Medicine* 36: 85–93.

Harwood, A., & Lakeman, G. 1995 We're known as the hen party because we lay eggs for sale. *Daily Mirror*, 1 November, pp. 6–7.

Horne, G., Hughes, S.M., Matson, P.L., Buck, P., & Lieberman, B. 1993 The recruitment of oocyte donors. *British Journal of Obstetrics and Gynaecology* 100: 877–8.

Human Fertilisation and Embryology Authority 1993 *Code of Practice*. HFEA: London.

Macbeth, H. 1993 Ethnicity and human biology, in Chapman, M. (ed.) *Social and Biological Aspects of Ethnicity*. Oxford University Press: Oxford.

Morgan, D., & Lee, R.G. 1991 *Blackstone's Guide to Human Fertilisation and Embryology Act 1990*. Blackstone: London.

Novaes, S. 1993 *Les Passeurs de gametes*. Presses Universitaires de Nancy: Nancy.

Peel, J. 1973 *Report of the Panel on Human Artificial Insemination* (Peel Report). Reprinted in *British Medical Journal* II (supplement): appendix V, 3.

Pfeffer, N. 1993 *The Stork and the Syringe*. Polity Press: Oxford.

Price, F. 1995 Conceiving relations: egg and sperm donation in assisted procreation, in Pearl, D., & Pickford, R. (eds) *Frontiers of Family Law*. Wiley: London, pp. 176–86.

Price, F., & Cook, R. 1995 The donor, the recipient and the child: human egg donation in UK licensed centres. *Child and Family Law Quarterly* 7: 1–7.

Royal College of Obstetricians and Gynaecologists 1983 *Report of the Ethics Committee on In Vitro Fertilisation and Embryo Replacement or Transfer*. RCOG: London.

Smith, B.H., & Cooke, I.D. 1991 Ovarian hyperstimulation: actual and theoretical risks. *British Medical Journal* 302: 127–8.

Timmins, N. 1994 Couples barred from choosing race of babies. *Independent*, 1 January.

Tizard, B., & Phoenix, A. 1993 *Black, White or Mixed Race?* Routledge: London.

Trounson, A.D., Leeton, J.F., Besanko, M., Wood, E.C., & Conti, A. 1983 Pregnancy established in an infertile patient after transfer of a donated embryo fertilised in vitro. *British Medical Journal* 286: 835–8.

Vines, G. 1994 Double standards for egg and sperm donors. *New Scientist*, 3 September, p. 8.

Warnock, M. 1984 *Report of the Committee of Inquiry into Human Fertilisation and Embryology*, Cmnd 9314. HMSO: London.

17 About Genetics: Aspects of Social Structure Worth Considering

Meg Stacey

The focus of this book – culture, kinship and genes – is of crucial importance. My own sense of this (Stacey 1988) has been confirmed by Marilyn Strathern's books (1992a, 1992b), the work that she led with colleagues in Manchester (Edwards et al. 1993) and her chapter in this volume. These contributions must have dispelled any lingering doubt a reader may have had of the subject's importance. In this chapter, I propose to complement the cultural focus by briefly drawing attention to structural aspects, for the applications of the new genetics have social as well as cultural implications. In doing this, I shall attempt to clear up some persistent misunderstandings about 'society' and 'the social'.

Interdisciplinary work is always difficult. The way anthropologists and sociologists identify problems, and the language they use to talk about them and their possible solutions, differ greatly from those used by professionals trained in clinical genetics and allied subjects. These differences apart, I take it that a common concern is to reduce suffering where we can.

When I talk about structure, I mean all those social arrangements which appear to be outside us, which exist before we arrive on the scene, which make social life possible and which provide 'the means, media, rules and resources for everything we do', as Bhaskar has put it (1989: 3–4, quoted in Porter 1993: 593). In one sense we inherit rather than create society: social structures were there before any one of us was born. However, neither 'society' nor social structures are static: they continually change, more or less rapidly. It is we who, in our everyday activities, continually reproduce or transform the structures. Thus, society cannot exist without us, that is to say outside human agency. In this sense, and in this sense only, Margaret Thatcher was right: 'society' does not exist. But society – social structures – are very much there as something, albeit abstract, which both enables and constrains our actions.

231

Medical practitioners seeking guidance as to what they should or should not do, may or may not be permitted to do, not infrequently speak of 'society' as if it were a being with a mind that it can make up. But society is not such a being. I suspect that what such practitioners are asking is that government should make up its mind. Government, or more accurately the members of the government, will do this by listening to opinion leaders, the media, constituents, civil servants and members of parliament. Many people, many human agents in many interactions, will be involved in a government conclusion about action or no action on, say, genetic population screening. But only limited sections of the total society will have been involved. The views that will prevail will be those of the elites, the dominant majority.

Given such complexities, unless they just want to pass the buck, practitioners and others should be clear about what they mean when they begin a sentence with 'society'. This casts society as an actor, but society cannot act – only humans can act. A society, the society, my or your society all make sense, but when I see 'Society', I do not know what it means. 'Society' may seem a useful shorthand, but the use can be dangerous; the reification lulls critical thought about who is doing, or being asked to do, what.

More does exist than 'individuals and their families' (the only realities Lady Thatcher was prepared to admit). Social structures do exist, but human agents continually reinforce or change them. Medical science and medical practice themselves constitute a part of those social structures. Through the agency of genetic screening and, later on, through genetic manipulation, the new genetics is making and will make changes in the social structures as well as in the meanings we put on our lives. To some extent, what changes are wrought will depend on how the procedures are applied, to what and to whom. There will be unintended consequences which may be beneficial or malign. Furthermore, the same action may not have the same effect on people throughout the society.

Many previous chapters have reflected structural issues, such as power relations and unequal resource distribution; structural issues may be expressed in formal institutions or more informally – science and technology have formal and informal structures; 'race' – a cultural concept – becomes institutionalised through racism; kinship arrangements, closely implicated in genetics, are part of the social structure. Behaviours and actions constitute the agency whereby the social structure is sustained or changed. The new genetics will undoubtedly have implications for structure as well as culture.

Some biologists share this view. Rose et al. (1984: 282) concluded on biological evidence, supported by the twitching muscle of the frog, that 'The biological and the social are neither separable, nor antithetical, nor alternatives, but complementary', thus dismissing the separate existences, or roles, of the organism and its environment.

THE MESSY NATURE OF SOCIAL REALITY

Genetics may be a rational science, but the organisation of our society is not rational, although that is not to say that one cannot use reason to examine it. There are many deep divisions, or fractures, in its fabric. Divisions as to material and social resources are profound. Social class is different, both in character and in where the fractures run, from earlier this century, but the divisions nevertheless remain deep. Differences in wealth are now greater than they were in 1970. Along with these go differences in health experiences. Phillimore et al. (1994) report that the latter increased in the decade 1981–91 in the North of England. The complex way in which health differentials are established is illustrated by Baker (1995). Access to health care is also not evenly distributed.

Naomi Pfeffer (1992, 1993) has pointed out how infertility treatments have always been differentially more available to the better-off, a trend which has increased with high-technology applications, such as in vitro fertilisation (IVF). The likelihood of such differences also applying in some genetic procedures must be high, accentuated by the continuing changes in the National Health Service. The possible implications of systematic variation, by some form of social class division, in the availability of genetics services are quite chilling. Could we perhaps unwittingly create an entirely new version of Disraeli's two nations, one in which human intervention in the human body is actively involved, the two nations created being differentiated by physical and mental ability – and thereby also differentiated in their access to and control over resources?

The Nuffield Report on the ethics of genetic screening speaks of the increased choices which would become available to persons who had learned that their genetic profile put them at risk in particular occupations (Nuffield Council on Bioethics 1993: para. 6.6). They would have the freedom to decide whether to risk the consequences or refuse the job. As I have noted elsewhere (Stacey 1994), one can hear the hollow laughter of the out-of-work: 'A chance would be a fine thing.' Unemployment rates are high; we are told this will continue. So much for rational choice.

SEX AND GENDER

Gender (as distinct from sex) constitutes another major structural division to which less attention has been paid in this volume and elsewhere than one might have expected (a more extended discussion is in Stacey 1996). I would have expected gender to rate higher, considering the variations in the relative roles of women and men in the gender order and in the tasks they perform according to the traditions of different religious and ethnic groups; it is also surprising since it is the minds and bodies of women which take the brunt of reproductive choices. Women were revealed as the 'kin keepers' in the early 1950s' studies of English kinship (Young & Willmott 1957, Firth 1956, Stacey 1960). Martin Richards (1996), has suggested that women are 'the genetic housekeepers'.

RACE AND ETHNICITY

Furthermore, ours (to my sorrow and shame) is a racist society. The racism and the failure to face its implications show up in many documents about the new genetics and genetic screening. The way questions to do with 'race' and ethnicity are dealt with or ignored when genetics are being discussed reveal our unpreparedness to handle the interface between the biological and the social. In the *Report of the Working Party of the Standing Medical Advisory Committee on Sickle Cell, Thalassaemia and other Haemoglobinopathies* (1993), the adjectives 'racial' and 'ethnic' are both used, sometimes together. The Nuffield Report on the ethics of genetic screening (1993: ch. 8), the reports of the Human Fertilisation and Embryology Authority (HFEA), and the Clothier Report on the ethics of gene therapy (1992) all duck, or skate round, the issues.

Others in this volume (and see also Bradby 1996) have discussed how certain genetic conditions are found more frequently in peoples of a particular descent and how ethnic groups are sometimes associated with peoples of a common descent. But not always. Any hint that the clusters of people we are talking about are 'races' with different diseases would be unwise as well as untrue. It is 50 years since biologists, in the wake of the Nazi horrors in which medicine was converted from a healing to a killing mission, sat down together and pointed out that 'race' was a meaningless concept. Social scientists have struggled to overcome the legacy of imperialism in their work. We now have to face the question: how are we to handle the uneven spread of genetic differences among the world's population, and apply the knowledge which genetic testing can give about disease conditions, without increasing racism and perhaps giving it new and subtler meanings? Genetic scientists and medical geneticists are in crucial positions both in what they say and do and in what they do not say.

ACTION AND INTERACTION – AN EXAMPLE

The Human Diversity Genome Project illustrates well the research hazards around supposed links between genetic dispositions and cultural groups (Lock 1994). Loosely affiliated to the Human Genome Organisation (HUGO), one objective is:

> to counteract in some measure the excessive standardization and parochialism associated with HUGO, while at the same time contributing to a further debunking of racial mythologies by showing the paramount importance of individual variation. (Lock 1994: 604)

However, problems arose. Academically, anthropologists were not happy with one geneticist's suggestion that a population would qualify for inclusion if it had been 'in place as of 1492' (Roberts 1992, cited in Lock 1994: 604); nor with the inclusion of the 'Etas' of Japan: 'This pejorative name, no longer used in Japan, designated an urban outcast group, which, despite marriage prohibitions, never remained biologically isolated' (Lock 1994: 605).

Doubts and fears were aroused in the 'indigenous' peoples. Not involved in the initial discussions, but well connected to contemporary communication systems, Iroquoian representatives protested by e-mail, angrily but with humour, when they learned the Iroquois were likely to be sampled (Lock 1994: 603, 605). As Lock (1994: 603) puts it, 'the native gaze, hypersensitive to exploitation (and no wonder), glower[ed] back'. RAFI (Rural Advancement Foundation International), which had originally alerted the Iroquois, was also bothered, among other things, about whether the six indigenous communities chosen for sampling in Iraq could be sure their cell lines stored in Baghdad 'will not be tampered with in some nefarious way which might lead to the destruction of the peoples themselves' – fears of a new kind of genocide (Lock 1994: 606).

Concluding her commentary, Lock points to possible advantages to be gained from the Diversity Project. For example, the genetic predisposition to diabetes, prevalent among North American indigenous peoples, did not manifest as disease until the mid-20th century; greater understanding would be beneficial if the knowledge were used 'not to create a genetically determined argument about diabetes' but to provide a better understanding of 'the relationship among culture, environment and genes' (Lock 1994: 606).

This account is not only an example of genetics as social action, but also demonstrates interactions showing the responses in action made by the recipients of actions initiated elsewhere. Also illustrated are the research hazards where supposed links between genetic dispositions and cultural groups are concerned. Research design is intricate where group characteristics have been socially constructed on the basis of alleged common genetic inheritance, especially given the close connection of many such constructions with remembered historical actions, such as genocide or colonisation. The intricacy also derives from the relative contemporary power of one group over another.

Avoiding issues such as these, as official reports do, is not the way to go about it – indeed, it is a good way of getting it wrong. Lock's account warns of ways in which the new genetics could become an ideology, geneticism, leading inadvertently – even among the overtly anti-racist – to racism. This need not come about: there are, after all, anti-racist genetic messages. These may miscarry, however, unless geneticists and the rest of us exhibit a sophisticated awareness to the complexity of the social issues involved, and unless we actively research them. The roots of racism lie deep in the social and cultural arrangements of our society. This volume represents an important beginning of the work which requires urgent extension: the sharing of social and biological knowledge about social inheritance and genetic inheritance and their interrelationships.

INTERPLAY OF STRUCTURES

Some of the most difficult problems emerge where a number of structures intersect, for example, 'race', gender and formal power, two of which, 'race' and gender, are social constructions based on apparent biological difference. TV images of assembled European leaders, almost exclusively white and male, demonstrate that those who hold political power are not representative of modern European populations. Government has expressed anxiety about the waste of resources implicit in the male domination of science. No similar governmental anxiety has yet been expressed about the monopoly of high places in politics or science by the dominant white ethnic group.

Some argue that the rational methodology of science overrides these biases, but such neutrality is absent in practice (e.g. Harding 1986). The different positions of women and men and of non-white and white in the social structure of the UK lead each to give priority to different problems, even to see them differently – not only in discovery but in the interpretation and, later on, in the application of the findings. Research suggests there may have been indirect discrimination against those of African descent in the identification of sickling disorders and in the service provision to handle resultant pain and illness (France-Dawson 1990; see also Green & France-Dawson, Chapter 8), and that the implication of cross-cousin marriages in poor obstetric outcomes have been misread (Ahmad 1994).

SCIENCE, TECHNOLOGY AND ESPECIALLY BIOLOGY

Our technological society has grown out of centuries of techno-scientific endeavour: its blessings and its curses are accepted with both gratitude and groans. Scientists are concerned that the public do not understand the methods or nature of the sciences on which all these developments, including genetics, are based. Maybe 'the public' do not understand science well, but it is held in high esteem, even if the relationship nowadays is a love–hate one (Stacey 1992). The different modes of understanding life and its chances that are espoused by scientists and the public express another structural division, one which leads, as many divisions do, to ranking.

The high ranking is expressed in yet another division, that between 'hard' or bench sciences and other systems of knowledge or science. The amount spent on research is one indicator of the institutionalised priority given to science and also to medicine. Table 17.1 shows, for 1991–92, the differential funding in rank order for each of the UK research councils. Of course, no council gets as much as it would like – and even less since public expenditure cuts have become such an overwhelming government concern. However, the data do show one structural effect of the kind of society we are in and the dominance of the scientific ideology.

Table 17.1 Government funding for UK research councils, 1991–92.

Research council	Funding (£m)
Science and Engineering (SERC)	443.3
Medical (MRC)	197.1
Natural Environment (NERC)	116.4
Agriculture and Food (AFRC)	86.1
Economic and Social Science (ESRC)	35.8

Source: *Annual Review of Government Funded Research and Development* (HMSO: London, 1993), Table 1.2.2.

When I complained at a CIBA meeting at the British Association in 1990 that the social sciences had been ignored in the development and application both of new reproductive technologies (NRTs) and of the new genetics, I was told I should have filled the gap and organised the social science research. 'But with what resources?' I said to myself, '... and how to get in there?' My attempts to persuade funders at the outset that they should also fund social research had been rebuffed (see Stacey 1992). However, times have changed. The Chalmers Report on technological applications in medicine (Department of Health 1992) says that health service and other social and cultural implications should be researched at an early stage.

Biologism

The biological sciences represent a particular facet of this scientific and technological dominance. Over the last 200 years, our society has gradually become biologistic and our medicine, biomedicine. Trust, even faith, was placed in the resultant techniques. Doubts about science and its applications in medicine have, however, begun to enter, in the latter case possibly beginning with the thalidomide tragedy (Stacey 1993).

IS COUNSELLING ENOUGH?

Counselling is an aspect of the 'social' which has been stressed by medical geneticists from the outset. The recent Nuffield Report reinforced this (1993: paras 4.17–4.22), indicating further that the careful and time-consuming counselling available in research centres is at risk of being lost in the stress of general service provision. The report goes on to explain how resources and training will be required for counsellors. It does not, however, suggest that the provision of adequate independent counselling should be a condition of the introduction of a programme, which, going by the working party's own arguments, it assuredly should be. In this way, the report colludes with the dominant biological ideology, which avers that our life courses and experiences are more influenced by our biology than by our social and cultural experiences.

NO IMPOSITIONS

The aim of ethically and democratically conscious geneticists is that there should be no impositions: individuals should be given the right to choose whether to know or not to know, whether to act on the information or not; screening and testing should be voluntary and, given proper information, decisions should be free as to what course an individual or a couple wishes to pursue.

In practice, given the social structures, such freedom is elusive if not illusory; the apparent freedom of choice could become another form of control. A normative consensus, once established, is hard to resist – the 'done thing', which 'everybody' does. The Nuffield Report is not the only one which implies, using rhesus and rubella screening as examples, that we need not worry any more because particular procedures 'seem to be well accepted' (para 3.34) (Stacey 1994, 1996). Maybe the outcomes justify the use of those procedures, but to state, as evidence of their worth, that there is no resistance to them is to confuse successful introduction with the ethical appropriateness of what was done. A horse which has been broken in and well trained no longer bucks and shies; it will now obediently serve its master. Given the structured inequality between doctor and patient and the trust which patients place in their practitioners, people can be persuaded to accept a wide range of procedures, especially if their use is relatively trouble free. Once established as routine, procedures tend to be accepted without question unless and until circumstances give rise to iconoclastic and radical questioning. (For a different approach to these problems see Raeburn 1991, 1994.) This underscores the importance of high-standard counselling offered by people who understand not only about genetics, but also about the contexts in which decisions are made – people who also are quite independent of those who will carry out the procedures and have no interest, material or other, in their use. A similar provision was suggested by the Polkinghorne Review (1989) on consent to the use of foetal tissue. Research has shown that those who undertake procedures unwittingly give advice in a way that is loaded to their point of view. Some may even effectively use blackmail. In 1980, Wendy Farrant asked 304 consultant obstetricians: 'Do you generally require that a patient should agree to the termination of an affected pregnancy before proceeding with amniocentesis/CVS?' Three quarters replied that they 'generally require women to agree to an abortion before proceeding to amniocentesis' (Farrant 1985: 113). In 1993, 357 obstetricians answered the same question put to them by Jo Green and her colleagues. One third of them answered 'yes' (Green et al. 1994, personal communication; Green & Statham 1996).

EUGENICISM: IT COULDN'T HAPPEN HERE

The Nuffield working party indicates (para. 8.19) that 'genetic testing in medicine in the UK is used to help individuals and their families avoid the occurrence of serious inherited disorders or their associated complications...[as are]...those wider population-based genetic screening programmes that have so far been established'. The report also points out that the UK has never had, despite some support for the idea, any eugenic legislation. The HFEA consultation document on sex selection calms fears on the grounds that legislation is in place (HFEA 1993: 24). But legislation can be changed, as have our civil rights since the early '80s.

How can we convincingly prevent any slide from a policy of helping individuals towards the goal of 'improving the quality of the population'? Paul Weindling's evidence about pre-Nazi Germany (1993) shows how a public hygiene programme prepared German doctors for the misuse of medicine in search of the pure German people.

Germany in the 1920s, 1930s and 1940s may have been a special case, as indeed it was, but we need to be ever-vigilant. In a libertarian society such as ours, given the power of normative control, an unremarked concatenation of decisions and arrangements could lead us into an effectively totalitarian situation. An important task is to ensure that this cannot happen. Detached and systematic investigation to understand the social processes involved in on-going developments is essential.

SCIENTISTS' UNDERSTANDING OF THE PUBLIC

The effort to increase public understanding of science is appropriate in a world so strongly shaped by science and technology. The teaching of genetics as part of the National Curriculum has been recommended: it would need to include social and psychological facets of inheritance as well as biological. Just as urgent – if not more so – is that scientists, and specifically geneticists, should come to understand the public. In Chapter 13, Martin Richards shows just how different is ordinary people's understanding of heredity from that of scientists, and that it can be systematically researched. Research is in train, but more is needed, especially given both the variations in the behaviour and experience of genetic conditions and the many divisions in our society (see the social research agenda proposed by Davison et al. 1994).

Knowledge may, of course, be used for many purposes. A scientistic – as opposed to a scientific – approach might be to use the knowledge the better to override popular ideas, to improve ways of breaking in the horse of public opinion. That route would lose much important diversity and many valuable insights and understandings. A scientific approach would more likely consider the implications of popular understanding for improved clinical care; more than that, it would examine its implications for the genetic research agenda itself.

Neither social nor medical sciences have all the answers: human values are also involved. As genetic screening becomes more common, popular understanding in geneticists' terms will increase. However, human values suggest the need for everyone to be free and able to retain or develop their own understandings of what inheritance may mean, what risks they are prepared to take, what life and death may mean, and what pain and suffering are about. That freedom may become increasingly difficult to retain, because the new genetics, like the NRTs, bring more third-party control with them. Effort will be needed on all our parts, not least the geneticists', to ensure that the freedom is preserved and enhanced.

CONCLUSION

Let me conclude by quoting one geneticist's view of the relation between the social and the genetic. Lewontin (1993: 123), in his 1990 Massey Lectures on the ideology of science, said that:

> the genes, in making possible the development of human consciousness, have surrendered their power both to determine the individual and its environment. They have been replaced by an entirely new level of causation, that of social interaction with its own laws and its own nature that can be understood and explored only through that unique form of experience, social action.

REFERENCES

Ahmad, W.I.U. 1994 Reflections on the consanguinity and birth outcome debate. *Journal of Public Health Medicine* 16: 423–8.

Baker, D. 1995 Poverty and disease: a postcard from the edge. *Journal of the Royal Society of Medicine* 88(3): 127–9.

Bhaskar, R. 1989 *Reclaiming Reality: A critical introduction to contemporary philosophy*. Verso: London.

Bradby, H. 1996 Genetics and racism, in Marteau, T., & Richards, M. (eds) *The Troubled Helix: Social and psychological implications of the new human genetics*. Cambridge University Press: Cambridge, pp. 295–316.

Clothier, C.M. 1992 *Report by the Commission on the Ethics of Gene Therapy*, Cm. 1788. HMSO: London.

Davison, C., Macintyre, S., & Smith, G.D. 1994 The potential social impact of predictive genetic testing for susceptibility to common chronic diseases: a review and proposed research agenda. *Sociology of Health and Illness* 16(3): 340–71.

Department of Health 1992 *Assessing the Effects of Health Technologies: Principles, practice, proposals* (Chalmers Report). Department of Health: London.

Edwards, J., Hirsch E., Franklin, S., Price, F., & Strathern, M. 1993 *Technologies of Procreation: Kinship in the age of assisted conception*. Manchester University Press: Manchester.

Farrant, W. 1985 Who's for Amniocentesis? The politics of prenatal screening, in Homans, H. (ed.) *The Sexual Politics of Reproduction*. Gower: Aldershot.

Firth, R. (ed.) 1956 *Two Studies of Kinship in London*. London School of Economics, Monograph in Social Anthropology 15. Athlone: London.

France-Dawson, M. 1990 *Sickle Cell Conditions – the Continuing Need for Comprehensive Health Care Services: A study of patients' views*. Daphne Heald Research Unit, Royal College of Nursing: London.

Green, J., & Statham, H. 1996 Psychosocial aspects of prenatal screening and diagnosis, in Marteau, T., & Richards, M. (eds) *The Troubled Helix: Social and psychological implications of the new human genetics*. Cambridge University Press: Cambridge.

Harding, S. 1986 *The Science Question in Feminism*. Open University Press: Milton Keynes.

Human Fertilisation and Embryology Authority 1993 *Sex Selection*. Public Consultation Document, HFEA: London.

Lewontin, R.C. 1993 (1991) *The Doctrine of DNA: Biology as ideology*. Penguin: Harmondsworth.

Lock, M. 1994 Interrogating the Human Diversity Genome Project. *Social Science and Medicine* 39: 603–6.

Nuffield Council on Bioethics 1993 *Genetic Screening: Ethical issues*. Nuffield Council on Bioethics: London.

Pfeffer, N. 1992 From private patients to privatization, in Stacey, M. (ed.) *Changing Human Reproduction: Social science perspectives*. Sage: London.

Pfeffer, N. 1993 *The Stork and the Syringe: A political history of reproductive medicine*. Polity Press: Cambridge.

Phillimore, P., Beattie, A., & Townsend, P. 1994 Widening inequality of health in northern England 1981–1991. *British Medical Journal* 308: 1125–8.

Polkinghorne Review 1989 *Review of the Guidance on the Use of Fetuses and Fetal Material*, Cm. 762. HMSO: London.

Porter, S. 1993 Critical realist ethnography: the case of racism and professionalism in a medical setting. *Sociology* 27(4): 591–609.

Raeburn, J.A. 1991 What do young people think about cystic fibrosis? *Journal of Medical Ethics* 28: 322–4.

Raeburn, J.A. 1994 Screening for carriers of cystic fibrosis. *British Medical Journal* 308: 1451–2.

Report of the Working Party of the Standing Medical Advisory Committee on Sickle Cell, Thalassaemia and other Haemoglobinopathies 1993. HMSO: London.

Richards, M. 1996 Families, kinship and genetics, in Marteau, T., & Richards, M. (eds) *The Troubled Helix: Social and psychological implications of the new human genetics*. Cambridge University Press: Cambridge, pp. 249–73.

Roberts, L. 1992 How to sample the world's genetic diversity. *Science* 257: 1204–5.

Rose, S., Lewontin, R.C., & Kamin, L.J. 1984 *Not in Our Genes*. Penguin: Harmondsworth.

Stacey, M. 1960 *Tradition and Change: A study of Banbury*. Oxford University Press: London.

244   *Culture, Kinship and Genes*

Stacey, M. 1988 The manipulation of the birth process: research implications. Paper presented to the February meeting of the WHO Advisory Committee for Health Research, Copenhagen.
Stacey, M. 1992 Social dimensions of assisted reproduction, in Stacey, M. (ed.) *Changing Human Reproduction: Social science perspectives*. Sage: London.
Stacey, M. 1993 The biological vision: triumphs and hazards. *Social Science and Medicine* 37(7): v–ix.
Stacey, M. 1994 Outlook of medical sociology: some sociological implications of genetic screening. Paper presented to the Genetic Screening Symposium, Centre of Medical Law and Ethics and Nuffield Council, Kings College London, March.
Stacey, M. 1996 The new genetics: a feminist view, in Marteau, T., & Richards, M., *The Troubled Helix: Social and psychological implications of the new human genetics*. Cambridge University Press: Cambridge.
Strathern, M. 1992a *After Nature: English kinship in the late twentieth century*. Cambridge University Press: Cambridge.
Strathern, M. 1992b *Reproducing the Future: Essays on anthropology, kinship and the new reproductive technologies*. Manchester University Press: Manchester.
Weindling, P. 1993 Medical degradation: the German medical profession 1900–1945. Paper presented at a meeting on Ethics in Health Care, Human Values in Health Care Forum, London, November.
Young, M., & Willmott, P. 1957 *Family and Kinship in East London*. Routledge and Kegan Paul: London.

18 Culture and Genetics: Is Genetics in Society or Society in Genetics?

Evelyn Parsons

> When I find myself in the company of scientists I feel like a shabby curate who has strayed by mistake into a drawing room of dukes.
>
> W.H. Auden

It was in 1962, during a row between F.R. Leavis and C.P. Snow, that the phrase 'two cultures' was used to distinguish what was defined as an 'unbridgeable' gulf between the grand dukes of science and the shabby curates of the arts. More recently, during a heated exchange at the British Association of Science (in 1994), Professors Francis Collins and Lewis Wolpert revisited the debate with the traditional arguments being made that art is too vague and it is scientists who really change the world. Science and 'art', or culture, have been perceived as non-intersecting, unrelated fields of study with science prevailing as the dominant paradigm. Dualism has been the traditional approach, science being associated with objective knowledge and the generation of 'facts' verified by empirical experiment, culture being relegated to the area of subjective experience. Capra argues that society, ever since Galileo, Descartes and Newton, has been so obsessed with rational knowledge, objectivity and quantification that we have become very insecure in dealing with human values (Capra 1982: 350).

The aim of this paper is to offer a sociological approach to understanding the relationship between culture and genetics. A review of the theoretical debates in the sociology of culture and in genetics will highlight that within both disciplines, two parallel paradigms have existed: one which is scientific and structural and one which is interactive. This paper will argue that an understanding of these two paradigms, with their polarised philosophies, provides the background to current debates about prenatal screening, presymptomatic testing and recent calls for greater government control of genetics. It will discuss the issue of whether the 1990s is a decade in which genetics is in society, or society is in genetics. It will ask whether we are experiencing the geneticisation of culture or the culturation of genetics. There are writers who would argue that modern genetics is still rooted in a dualistic, scientific, 'good gene' vs. 'bad gene' dichotomy which takes little account of social and cultural influences, and that it is this approach which informs policy decisions about screening programmes. In their view, we are experiencing the geneticisation of culture. Others see the practice of genetics as less conspiratorial, with a shared discourse at both the micro level, between individuals and professionals, and at the macro level between culture and genetics.

SOCIOLOGY OF CULTURE: THEORETICAL APPROACHES

Attempts to understand the complexities of culture, how it comes into being and then operates in everyday life, have occupied both anthropologists and sociologists for many years. It has been said that culture is difficult to talk about and even more difficult to define. Its very definition depends on whether you adopt a structural approach to the study of society or a social interactionist one. Structural writers distinguish between the subjective world of culture and the structural world of social systems which, they argue, have objective facticity and can be studied using the scientific method. Weber (1963) referred to culture as a matter of ethic or spirit, something quite different from other social arrangements. Early anthropologists, accepting this dualism, defined culture in terms of a 'reality' which existed outside the individual and was acquired through a process of socialisation. For Tylor (1924), 'culture is that complex whole acquired by man as a member of society'. From this perspective, culture exercises a degree of social control and constraint through commonly held norms and values in society. This social control ensures social order and stability.

In contrast, the social interactionist perspective makes no distinction between social processes and social systems. Culture is seen to be constructed by individuals in a continuous process of interaction with other individuals and other social systems. All human societies are interpenetrated with culture, and cultural systems are articulated in social systems.

One common theme in the sociology of culture is the recognition that culture provides social order, whether imposed externally or constructed individually. Without culture, society would not exist. At the micro level, culture provides individuals with meaning, and at the macro level, patterns of meaning are embodied in cultural symbols to provide cohesion.

The relationship between the two themes of individual meaning and cultural symbols are well illustrated in the work of Peter Berger. Berger (1967) argues that the ultimate terror for any individual would be to live in a world without structure or meaning. Culture makes the social world meaningful and provides protection against chaos. Individuals strive to construct meanings and are therefore the 'culture-makers'; they create society in a continual process of social exchange. Culture is not a 'once and for all' given reality; rather, it is constantly being constructed and reconstructed. Berger does concede that, at times, culture appears to be 'real', not because it is an objective, external reality but because it has become subjectively meaningful, something which is confirmed during social interaction with others. Situations do arise when individuals feel that culture is an external product, but this is because 'man merely forgets that the world he lives in has been produced by himself'.

Central to the human world is socially constructed meaning. Individuals, of necessity, search for meaning in situations and infuse it into their everyday reality. Subjective meanings become attached to individuals' own social actions, the social actions of others and social systems. Berger argues that these meanings become reabsorbed into consciousness as subjectively plausible definitions of reality and generate recipes for everyday living. Order is possible at the individual level because of subjectively constructed meanings, and at the level of society through the collective participation in symbols, symbolic systems and institutionalisation. Any society has an overarching organisation of symbols which have been generated and have become meaningful to individuals and consequently legitimated. A symbolic system for Berger is any significant theme that spans spheres of reality; he talks about the spheres of religion, art, philosophy, politics, psychology, etc. Symbolic representations impose themselves on everyday life by providing meaning. They legitimate activity at both the individual and the societal level. Symbols are individually constructed and therefore vary over time because they reflect changing definitions in society. For example, at one point in time, religion may be accepted as a symbolic system which provides individuals with meaning, whilst at another time, science, medicine or even medical specialities such as genetics may fulfil that symbolic role. For Berger, there is a constant dialectic between individuals and social systems, and the key element in the interaction is the exchange of knowledge. Knowledge is shared at the levels of the individual, the group and society as a whole, and thereby – as with symbolic systems – specific types of knowledge become legitimated.

The three themes of individual meaning, symbolic systems with their powers of legitimation, and the on-going dialectic between individuals and social systems in the construction of culture and everyday reality will be discussed later in the context of the 'culturation of genetics' debate. The next section, however, will briefly explore two perspectives on the relationship between culture and genetics.

GENETICS AND CULTURE: THEORETICAL APPROACHES IŃ GENETICS

Genetics is one field of science where the 'two culture' debate is frequently revisited. The crucial question that has vexed the academic world for more than a century is whether the mystery of life lies in nature or nurture, in an individual's DNA or in society. Since the mid-1980s, billions of dollars have been spent mapping the human genome with the intention of providing medical research with a full blueprint of human biology. But however successful this cartography might appear, it still fails to answer fundamental questions about the very nature of our existence. Weber, in his treatise on science as a vocation, argues that scientific progress represents a fraction of the overall process of intellectualisation, and it is naive to believe that life will be mastered by the techniques of science:

> Science is meaningless because it gives no answers to the only question important for us: what shall we do and how shall we live? (Weber, in Gerth & Wright Mills 1948: 143)
> Consider modern medicine which is highly developed scientifically.... Natural science gives us an answer to the question of what we must do if we wish to master life technically. It leaves quite aside... whether we do wish to... and whether it ultimately makes sense to do so. (Weber, in Gerth & Wright Mills 1948: 144)

Dualism in classical genetics was clear and simple: it was possible to identify 'good' genes and 'bad' genes, and the ills of humankind were seen to be the result of defective genes, not defective social or cultural environments. It was argued that a growing understanding of biology, and genetics in particular, would enable scientists to perfect human-kind:

> Science in the form of sociology and psychology will discover the qualities that are desirable for the best enjoyment of life, and science in the form of physiology and genetics... will discover what the elementary basis for these qualities are and will procure them for men. (H.J. Muller 1910, quoted in Wills 1993: 168 with amended punctuation)

The logical conclusion was self-evident: so-called 'good' genes should be promoted whilst 'bad' genes should be eliminated. This was the philosophy of eugenics (literally meaning well-born), a concept introduced by Galton in 1883 and one that has appeared in a number of guises since. East (1927) warned that civilisation was facing a dangerous situation which only the power of the (then) new genetics could avert by reducing the survival of the unfit. Genetics could claim to have 'noble aspirations' because it was helping the human race to dominate nature and keep the 'flock in prime condition'. The fundamental requirement for a sound and healthy body was a 'proper ancestry':

> We may react against a statement which assigns to free will no part in life, we may poke fun at what some have called 'Calvinistic predestination in scientific guise' but the facts remain. Let us think for a moment before we scoffingly pass judgement. Is a feeble-minded child likely to become a scholar?... Their heritage circumscribes their world. Is it so strange that heredity should set bounds for each one of us beyond which we cannot pass? (East 1927: 15)

More recently, sociobiologists have revived the debate, claiming that human history is ruled by our genes (Wilson 1975, Lumsden & Wilson 1981). A similar biological reductionism is reflected in Dawkins' work (1989) with his contention that life revolves around one element: competition to pass on one's genes. Genes, it is argued, not only determine phenotype in terms of hair colour, etc., but provide explanations for social and cultural behaviour, such as intelligence, criminal behaviour, sexual orientation, wealth, alcoholism, depression and dyslexia. The influence of culture is seen to be minimal; consider the role of Caliban in Shakespeare's *The Tempest*, or the arguments proposed in the more recent American publication, *The Bell Curve*, by Herrnstein and Murray (1994). Herrnstein and Murray claim that IQ is predetermined and predetermining, and that members of the underclass (mainly black) are therefore genetically preordained to be failures in Western society. Similarly, Beardmore and Karimi-Booshehri (1983), on the basis of linkage studies, claim that there is a significant correlation between blood group genes and social success in certain areas of England.

Clear parallels can be drawn between the classical, positivist traditions in both genetics and social science. Early sociologists claimed that their newly established discipline would make a major contribution to society. They argued that their science of society would enable them to predict human behaviour. Their precept was: To Know (individual social action) is To Predict (future individual patterns) is To Control (patterns of behaviour in society). This has since been echoed in the development of genetic techniques for predicting individuals' future health and life chances: To Know (individual DNA and genetic make-up) is To Predict (future health and susceptibility to disease by presymptomatic testing and prenatal screening) is To Control (development of disease through modification of lifestyle). For some, the answer to social ills lies in 'responsible' social and genetic engineering.

The classical and more recent socio-biological approaches have been challenged by both biologists and social scientists on the grounds that they have ignored one of humanity's most distinctive features: culture. Behaviour learned by an individual can be passed from one generation to another in a manner that is neither Darwinian nor genetic. Rose, Lewontin and Kamin (1984) challenge the very foundations of biological determinism that locates the causes of social phenomena in the genetic make-up of the individual. They argue that this deterministic perspective is being used as a powerful weapon to legitimate inequalities in society. Wilkie (1995a) reports a conversation with Roy Porter, the social historian of science and medicine, who said:

> If we believe that by understanding the genetic alphabet we'll understand better how society should be organised we're deluding ourselves... the unemployment figures tell us more than genetics about social behaviour. It's the cultural differences that are more interesting than the basic genetic background. We have the same genes as people had 30 years ago but our society is more violent. (Porter, quoted in Wilkie 1995a)

For Chasin (1980), any theories which assert fixed human qualities are political theories which are being used to justify situations of inequality. Clearly Wilson's argument (1975) that social stratification is a universal feature of human society, and that the wealthiest are only being rewarded for being the best, would fall into this category. Chasin says: 'Wilson, like other biological determinists, is wrong. The fault is in our society, the injustices, the inequalities do not lie in our genes' (p. 45). Duster (1990), in *Backdoor to Eugenics*, discusses both population screening and prenatal screening in the context of inequalities in society and, in particular, racial inequality:

Technical complexities of vanguard research in molecular biology and the promise of success incline us to go limp before such scientific know-how. We give up to 'science and expertise' a deep human concern that has little to do with either science or expertise, namely, what kind of knowledge we should pursue in determining what kind of children should be born. (p. viii)

For Duster, hysterical warnings of genocide will always fall on deaf ears, as they did in Germany in 1937 and America in 1930. The new attitude that a defective foetus can be eliminated is the present-day backdoor to eugenics.

The establishment in 1991 of the Human Genome Diversity Project is a reflection of concern about what is seen as the excessive standardisation being imposed by the Human Genome Programme. Lock (1994) points to a central paradox in genetics: whilst we share most of our genetic material in common with other members of our species, each individual being is genetically unique and is the repository of human genetic variation. Is it possible that the very element excluded from the majority of classical geneticists' formulations, that of culture, holds the key? It was the scientist Dobzhansky (1937) who argued that genes did not act alone but were in some form of complex interrelationship: 'Man is not just an overgrown Drosophila (fly). Some laws of biology apply however to men as well as to flies.'

This theme has been developed by Wills (1993), who suggests that the relationship between society and culture is not one of either total isolation or total absorption but one of constant feedback. This argument is very similar to the social interactionist perspective of Berger in his sociology of culture.

CULTURE AND GENETICS, OR GENETICS AND CULTURE?

The academic debate about the relationship between genetics and culture revolves around the issue of which element is defined as being the more powerful: does genetics impose itself on society, or does culture shape genetics in a process of interaction? Strong arguments have been put forward for both approaches.

The term 'geneticisation' has its roots in Merton's concept of scientisation (Stehr 1990) and medicalisation (Zola 1972, Illich 1976). Zola (1972), in his radical critique, claimed that medicine was becoming the new 'repository of truth' and, as such, was an institution of social control. Four years later, Illich argued that medicine is iatrogenic, having created a dependency not only at the individual level but at the social and cultural level:

> In medicine this damage appears as iatrogenesis. Iatrogenesis is clinical when pain, sickness and death result from medical care; it is social when health policies reinforce an individual organisation that generates ill-health; it is cultural and symbolic when medically sponsored behaviour and delusions restrict the vital autonomy of people, undermining their competence.... (Illich 1976: 270)

Awe-inspiring medical technology has combined with egalitarian rhetoric to create an illusion. In the words of Kennedy:

> The nature of modern medicine is positively deleterious to the health and well-being of the population. We have all been willing participants in allowing the creation of a myth...that health can be achieved, illness can be vanquished and death postponed until further notice...we all of us have hitched our wagon to the wrong star, scientific medicine. (Kennedy 1983: 25)

More specifically, Lippman (1992) argues that genetics has increasingly become the dominant discourse to explain health and disease, normality and abnormality. Two messages are being given: genes are causal, but genetics can be regulated. This process of geneticisation individualises responsibility for health and, in doing so, reinforces inequality by diverting attention away from social, physical and environmental factors. For Lippman, the Human Genome Project is a reflection of the genetic 'colonisation' of society; she argues that funding is only available for the 'mapmakers' because the vision of society is now filtered through a genetic prism, or, as Berger would argue, genetics has become a symbolic system which is accepted within society as a legitimate tool of explanation. This represents a sad commentary on a society which prefers to define relationships in terms of DNA rather than social relationships. For Lippman (1993a), biology is *not* destiny and science is *not* the only legitimate source of knowledge.

Rose (1987) makes a similar radical critique of genetics, arguing that science does not just compete with nature but seeks to dominate and exploit it. The explosive growth of molecular biology since the 1970s has meant that genetic information, once confined to the realm of scientific knowledge, has moved into the public domain with the potential for clinical applications. Any debate about genetics, for Rose, is a political debate that needs to be put in the context of the biological determinism of modern science. One area of particular concern for Rose and other feminist writers is prenatal screening.

For Farrant (1985), the offer of prenatal screening reinforces a dominant ideology that poses solutions to disability in terms of medical science and maternal responsibility rather than social and political change. Browner and Press (1995) argue, from their Canadian study, that alpha-fetoprotein (AFP) screening in pregnancy has become socially and culturally acceptable to the majority of women because the language used to promote the programme 'obscures its eugenic potential' (p. 308). They refer to programmes of prenatal screening as 'contemporary eugenic control programmes'. These 'neo-eugenic' programmes are designed to select out those with specific physical or mental disabilities, and rest on two assumptions which have permeated modern society: first, that scientific knowledge represents absolute truth; second, that as a culture we are preoccupied with the notion of risk and, in particular, with risk in pregnancy. In an earlier article, Press and Browner (1994) argue that AFP has become incorporated into culture because it has been seen as an uncomplicated and routine part of prenatal care, whilst there has been silence about other issues, such as the appropriateness of aborting foetuses with physical abnormalities.

There are two significant features about those who support a geneticisation thesis. First, while they maintain that genetics has been accepted as a symbolic system in society, they also maintain that it poses a major threat; its ability to define 'good' and 'bad' genes leads it towards defining what is normal (acceptable) and abnormal (unacceptable) in society. Second, genetics is seen as an external reality that has the power to exploit less powerful groups in society (particularly women) and to impose constraint on their freedom of choice. Farrant (1985) doubts the authenticity of the claim that advances in reproductive technology herald increasing choice for women:

> The concept of 'choice' about amniocentesis... needs to be located within the context of antenatal care which generally provides women with very little control over decisions about their own treatment. (p. 113)

Rothman (1987) similarly argues that there is only choice for those who want what society wants: 'The new technology... offers new choices but it also creates new structures and new limitations on choice' (p. 14).

The new genetics brings with it new levels of knowledge and awareness, which in turn infiltrate and change definitions in society as to the nature of responsible parenthood:

> Blame begins to insinuate itself. The birth of a severely disabled child, when the disability could have been prenatally diagnosed and the pregnancy terminated, begins to be seen as an act of irresponsibility. (Rothman 1987: 227)

The radical critique, centring on the concept of geneticisation, has been deemed far too extreme for many writers, both within and outside genetics. There is an awareness of the potential dangers from the new genetics, but they are deemed to be minimal. Harper (1988) admits that the potential for the abuse of genetics is now greater than ever before because technology has become much more powerful. He sees the major threat coming from the misuse of genetic registers and predictive testing rather than genetic manipulation. His response is more conservative than those who fear the colonisation of culture by genetics. He proposes that we maintain an open and free society with persistent vigilance, and that both scientists and clinicians should not place their work above the benefits to the families concerned. A similar confidence that many fears are ill founded is expressed in a lead article in *The Times* (Anon. 1994). Referring to the human genome project as 'leading to greater glories' (terminology far removed from that used by radical writers), the author says, 'in spite of the warnings of science fiction, the risk of a new authoritarianism arising from genetic technology still seems comparatively slight'. This contrasts sharply with

Lippman's warning (1993a, 1993b) that creating life forms is no longer pure fantasy and that, although we might not have to battle with the creatures of Jurassic Park, we must confront the possibly disastrous effects of genetic engineering: 'it is not just that there is a danger in tinkering – it is just as dangerous to think that we can do it better the next time'.

The conservative response to the development of the new genetics is reflected in calls for more interaction between society and science so that there is scrutiny from outside. The recent calls for the setting up of a human genetics commission (Wilkie 1995b, Clothier 1992, Nuffield Council on Bioethics 1993, House of Commons Select Committee on Science and Technology 1995) all reflect an increasing discourse and shared understanding between genetics and society. Macintyre (1995), in her discussion of the social context of the new genetics, calls for a dialectic in which there is not just a public understanding of science but also a scientific understanding of the public. This greater understanding of the public, she contends, will come from studies of what individuals, families and social institutions feel and do in relation to the new genetics. The underlying assumption, from this perspective, is that the new genetics can be harnessed for the good of society, by society, in a shared discourse.

It is this very area of psychosocial research in genetics that raises the most interesting element in the 'culturation of genetics' debate. Harper (1993) heralds psychosocial genetics as an emerging scientific discipline which will 'perhaps be the best safeguard against abuse' as trained, critical and involved social scientists work with clinical geneticists. Surely this is a valid argument: if social scientists are working with geneticists on a daily basis, they will be the first to 'blow the whistle' on so-called 'neo-eugenics' if it is present? The psychosocial research on genetic counselling and related topics – the relationship between risk and reproductive patterning, attitudes to population screening, experience of presymptomatic testing – is extensive and cannot be reviewed here save to highlight two issues. First, there is great stress laid on the social and cultural variables that intervene between genes and genetic decision-making. For example, mathematical levels of risk are not seen to be the sole determinant in reproductive patterning (Lippman-Hand & Fraser 1979a, 1979b, Wertz et al. 1986, Frets et al. 1990, Parsons & Atkinson 1992, 1993 Parsons & Clarke 1993), and the uptake of population screening programmes is not solely related to the nature of the test that is being offered but, more importantly, to the way in which the offer is made (Marteau 1989). Similarly, personal biographical factors are seen to play an important part in the uptake of presymptomatic testing (Tyler et al. 1992, Tibben et al. 1992, Huggins et al. 1992, Bloch et al. 1992, Holloway et al. 1994). Second, the majority of studies are specific and reflect research at the micro level; no overarching structural interpretation has been imposed.

The tension between the reporting of research at the micro level and the interpretation of those data at a structural level is illustrated in the work of Bosk (1992). Bosk, a social ethnographer, studied genetic counselling in an American paediatric hospital. At the micro level, he talks about genetic counsellors as a group of professionals with a complex knowledge base who have virtually no power or autonomy: 'They are information specialists, neutral staters of odds, . . . decision facilitators rather than decision makers' (p. xv). Bosk talks about the 'terrible ordinariness' of genetic counselling and their 'dreary work routine'. He then faces the question of how individual genetic counsellors (micro level) fit into the overall process of genetics at the structural (macro) level:

> Watching the counsellors at work it is hard to make any connection at all between past abuses of genetics and the practice of genetic counselling I observed. . . . In all medical practice it is hard to imagine a more prosaic activity performed with better intention than genetic counselling. (pp. xviii–xix)

Bosk then refers to the human genome mapping programme, prenatal screening and genetic engineering as 'audacious exercises'. For him, genetic counsellors are 'good people engaged in dirty work' (p. xxi). Here is the concept of innocent individuals being used by a society whose agenda is far more conspiratorial. The sum of the whole (that is, society as an external reality) is greater than the sum of its individual parts (its members). For Bosk, there was no evidence of conspiracy at the individual level, but he felt that there was something disturbing about the way genetics is applied (p. xix). Rapp, in her study on the discourse of genetic counselling (1988), refers to counsellors as 'heirs to a eugenic script'. However much they may aspire to liberate women and children, they are confronted by the contextual setting of their work. For Rapp, genetics has the potential to be both emancipatory and socially controlling.

CONCLUSION

The chasm between those who fear that geneticisation is a reality and those who accept the incorporation of genetics into culture as a positive advance seems unbridgeable. For those who accept the geneticisation argument, there is, in effect, a conspiracy for genetics to colonise society, something of which individuals and health professionals are largely unaware as they practice the science of genetics and the art of medicine. The role of the social scientist is therefore to highlight these structural issues and to raise individual and societal awareness. For those who favour the culturation argument, no such conspiracy exists; individuals have 'merely forgotten that the social world they live in has been constructed by themselves', as Berger would argue.

There has been concern about the new genetic 'prism', but there is an equally powerful second prism, the social and cultural one. It is this second prism which influences social scientists in their interpretation of what is going on in society, at both the macro and micro levels. We are in a period when the cultural prism has never had greater access or greater freedom to interpret genetic happenings. One important lesson one needs to learn is that no social scientist is value free. To identify a social scientist's perspective provides the key to understanding the framework they will use to interpret their data, and this allows the reader to develop their own personal understanding of the relationship between genetics and culture.

REFERENCES

Anon. 1994 The gene genie: the Human Genome Project should lead to greater glories. *The Times*, 2 May.
Beardmore, J.A., & Karimi-Booshehri, F. 1983 ABO genes are differentially distributed in socio-economic groups in England. *Nature* 303: 522–4.
Berger, P., with Luckmann, T. 1967 *The Social Construction of Reality: A treatise in the sociology of knowledge.* Penguin: Harmondsworth.
Bloch, M., Adam, S., Wiggins, S., Huggins, M., & Hayden, M.R. 1992 Predictive testing for Huntington disease in Canada: the experience of those receiving an increased risk. *American Journal of Medical Genetics* 42: 499–507.
Bosk, C.L. 1992 *All God's Mistakes.* University of Chicago Press: Chicago.
Browner, C.H., & Press, N.A. 1995 The normalization of prenatal diagnostic screening, in Ginsburg, F.D. & Rapp, R. (eds) *Conceiving the New World*

Order: The global politics of reproduction. University of California Press: Berkeley.

Capra, F. 1982 *The Turning Point: Science, society and the rising culture*. Flamingo: London.

Chasin, B. 1980 Sociobiology, a pseudo-scientific synthesis, in Arditti, R., Brennan, P. & Cavrak, S. (eds) *Science and Liberation*. South End Press: Boston.

Clothier, C.M. 1992 *Report by the Commission on the Ethics of Gene Therapy*, Cm. 1788. HMSO: London.

Dawkins, S. 1989 *The Selfish Gene*, 2nd edition. Oxford University Press: Oxford.

Dobzhansky, T. 1937 *Genetics and the Origin of Species*. Columbia University Press: New York.

Duster, T. 1990 *Backdoor to Eugenics*. Routledge: New York.

East, E.M. 1927 *Heredity and Human Affairs*. Charles Scribner: New York.

Farrant, W. 1985 Who's for amniocentesis? The politics of prenatal screening, in Homans, H. (ed.) *Sexual Politics of Reproduction*. Gower: Vermont.

Frets, P.G., Duivenvoorden, H.J., Verhage, F., Niermeijer, M.F., van de Berge, S.M.M., & Galjaard, H. 1990 Factors influencing the reproductive decision after genetic counseling. *American Journal of Medical Genetics* 35: 496–502.

Gerth, H.H., & Wright Mills, C. (eds) 1948 *From Max Weber: Essays in sociology*. Routledge: London.

Harper, P.S. 1988 The abuse of genetics: lessons from the past, lessons for the future. Editorial, *Journal of Medical Genetics* 25: 793.

Harper, P.S. 1993 Psychosocial genetics: an emerging scientific discipline. Editorial, *Journal of Medical Genetics* 30: 537.

Herrnstein, R.J., & Murray, C.J. 1994 *The Bell Curve: Intelligence and class structure in American life*. Free Press: New York.

Holloway, S., Mennie, M., Crosbie, A., Smith, B., Raeburn, S., Dinwoodie, D., Wright, A., May, H., Calder, K., Barron, L., & Brick, D.J.H. 1994 Predictive testing for Huntington disease: social characteristics and knowledge of applicants' attitudes to the test procedure and decisions made after testing. *Clinical Genetics* 46: 175–80.

House of Commons Select Committee on Science and Technology 1995 *Human Genetics: The science and its consequences*. HMSO: London.

Huggins, M., Bloch, M., Wiggins, S., Adam, S., Suchowersky, O., Trew, M., Klimek, M., Greenberg, C.R., Eleff, M., Thompson, L.P., Knight, J., MacLeod, P., Girard, K., Theilmann, J., Hedrick, A., & Hayden, M.R. 1992 Predictive testing for Huntington disease in Canada: adverse effects and unexpected results in those receiving a decreased risk. *American Journal of Medical Genetics* 42: 508–15.

Illich, I. 1976 *Limits to Medicine. Medical Nemesis: The expropriation of health*. Marion Boyars: London.

Kennedy, I. 1983 *The Unmasking of Medicine*. Granada: London.

Lippman, A. 1992 Led (astray) by genetic maps: the cartography of the human genome and health care. *Social Science and Medicine* 35(12): 1469–76.

Lippman, A. 1993a Jurassic Park opens gates to public discussion. *GeneWatch* 9(1–2): 4–5.

Lippman, A. 1993b Genetic engineering: cause for caution. *Globe and Mail*, 25 June.
Lippman-Hand, A., & Fraser, C. 1979a Genetic counseling: provision and reception of information. *American Journal of Medical Genetics* 3(2): 113–27.
Lippman-Hand, A., & Fraser, C. 1979b Genetic counseling – the postcounseling period: parents' perceptions of uncertainty. *American Journal of Medical Genetics* 4(1): 51–71.
Lock, M. 1994 Interrogating the human diversity genome project. Editorial, *Social Science and Medicine* 39(5): 603–6.
Lumsden, C.J., & Wilson, E.O. 1981 *Genes, Mind and Culture: The co-evolutionary process*. Harvard University Press: Cambridge, Mass.
Macintyre, S. 1995 The public understanding of science or the scientific understanding of the public? A review of the social context of the 'new genetics'. *Public Understanding of Science* 4(3): 223–32.
Marteau, T. 1989 Framing of information: its influence upon decisions of doctors and patients. *British Journal of Psychology* 28: 89–94.
Nuffield Council on Bioethics 1993 *Genetic Screening: Ethical issues*. Nuffield Council: London.
Parsons, E.P., & Atkinson, P.A. 1992 Lay constructions of genetic risk. *Sociology of Health and Illness* 14: 437–55.
Parsons, E.P., & Atkinson, P.A. 1993 Genetic risk and reproduction. *Sociological Review* 41(4): 679–706.
Parsons, E.P., & Clarke, A.J. 1993 Genetic risk: women's understandings of carrier risks in Duchenne muscular dystrophy. *Journal of Medical Genetics* 30: 562–6.
Press, N.A., & Browner, C.H. 1994 Collective silences, collective fictions: how prenatal diagnostic testing became part of routine prenatal care, in Rothenberg, K.H., & Thomson, E.J. (eds) *Women and Prenatal Testing: Facing the challenges of genetic technology*. Ohio State University Press: Columbus.
Rapp, R. 1988 Chromosomes and communication: the discourse of genetic counseling. *Medical Anthropology Quarterly* 2(2): 143–57.
Rose, H. 1987 Victorian values in the test-tube: the politics of reproductive science and technology. In Stanworth, M. (ed.) *Reproductive Technologies*. Polity: Cambridge.
Rose, S., Lewontin, R.C. & Kamin, L. 1984 *Not in Our Genes: Biology, ideology and human nature*. Penguin: London.
Rothman, B.K. 1987 *The Tentative Pregnancy: Prenatal diagnosis and the future of motherhood*. Penguin: New York.
Stehr, N. 1990 Robert K. Merton's sociology of science, in Clark, J., Modgil, C., & Modgil, S. (eds) *Robert K. Merton: Consensus or controversy*. Falmer Press: London.
Tibben, A., van der Vilis, M.V., Skraastad, M.I., Frets, P.G., van der Kamp, J.J.P., Niermeijer, M.F., van Ommen, G.J.B., Roos, R.A.C., Rooijmans, H.G.M., Stronks, D., & Verhage, F. 1992 DNA-testing for Huntington's disease in the Netherlands: a retrospective study on psychosocial effects. *American Journal of Medical Genetics* 44: 94–9.
Tyler, A., Ball, D., Craufurd, D. 1992 Presymptomatic testing for Huntington's disease in the United Kingdom. *British Medical Journal* 304: 1593–6.

Tylor, E.B. 1924 *Primitive Culture.* Bretano's: New York.

Weber, M. 1963 *The Sociology of Religion.* Beacon: Boston.

Wertz, D.C., Sorenson, J.R., & Heeren, T.C. 1986 Clients' interpretation of risks provided in genetic counseling. *American Journal of Human Genetics* 39: 253–64.

Wilkie, T. 1995a Science confronts the beast within. *Independent*, section 2, 17 July, pp. 2–3.

Wilkie, T. 1995b MPs warn against abuse of human genetics. *Independent*, 19 July, p. 7.

Wills, C. 1993 *The Runaway Brain: The evolution of human uniqueness.* Harper Collins: London.

Wilson, E.O. 1975 *Sociobiology: The new synthesis.* Harvard University Press: Cambridge, Mass.

Zola, E. 1972 Medicine as an institution of social control. *Sociological Review* 20: 487–504.

Index

ethnocentrism, 16
eugenics, 240–1, 249, 251, 253, 256
    and learning difficulties, 207, 209,
        210, 214, 216
European kinship patterns, 29–31
euthanasia, 150
Evans, D.G.R., 185, 186
Evans-Pritchard, E., 68
exclusion testing, Huntington's
    disease, 178, 180–1
extended families
    black population, Southern Africa,
        153
    consanguinity, 88–9, 116
    cultural understanding, 112–13,
        116, 117
    external locus of control, 148

familiarity with diseases, pregnant
    women, 125
family dynamics, 112–13
family planning, 154
fantasy, unconscious, 158–64
Farrant, Wendy, 198, 240, 253, 254
fascism, 215
fatalism, 148–9
'faulty genes', 7–8
Federation of Sickle Cell Clubs of
    Nigeria, 141
Feversham Committee, 219
first-cousin marriage
    black population, Southern Africa,
        151
    Mediterranean kinship patterns, 30
    Northern European kinship
        patterns, 30
    Pakistanis, British, 28, 76, 87, 101,
        106
    patrilineal kinship patterns, 34–7
    social structure, 237
foetal sex selection, 36, 42
Frankenberg, Ruth, 228
fraternal *biradheri*, 91
Freud, Sigmund, 158

Galton, Sir Francis, 249
gamete donation, 218–19, 228
    anonymity and the origins of
        'matching', 221–2

background, 219–21
    culture seen as barrier to scientific
        understanding, 46, 47, 51–2n.5
    'ethnic' donors, 225–8
    genetic screening of donors and
        recipients, 222–5
    research on embryos from, 52n.11
Gazankulu, 154
gender differences, 168, 234
gene clustering, 60–1, 62
gene frequencies
    and ancestral 'purity', 64
    clinical distributions, 57, 59, 60–1,
        62
    divergence, 55
general practitioners (GPs)
    Asian, 124, 129
    pregnant women, 122, 124
genes
    'faulty', 7–8
    language issues, 153, 190
    nature of, 5–7
genetic counselling, 8–10
    anthropological perspective on
        culture, 40
    black population, Southern Africa,
        152, 153, 155
    breast and ovarian cancer, 185, 187
    consanguinity, 116
    and cultural difference, 3
    cultural understanding, 117
    ethics of relationships, 200–1
    ethnic groups, 77
    haemoglobin disorders, 28
    Huntington's disease, 179, 180
    importance, 111, 240
    kinship, 27
    and lay knowledge, 188, 191
    limitations, 239
    micro and macro level, 256
    probability and odds calculations,
        communication of, 172
    sickle cell anaemia, 141–3
    social studies, importance, 15–16,
        17–18
    thalassaemia, 85, 91–5
    unconscious fantasy, role of,
        158–64
    vocabulary, 8

linkage testing
  breast and ovarian cancer, 182,
    183, 192
  Huntington's disease, 177, 178,
    179, 191
Lippman, A., 252–3, 255
Livingstone, D., 150
Lock, M., 235, 236, 251
Locke, John, 212
low birth weight, 98, 99, 107
Luther, Martin, 212

Mabayoje, J.O., 135–6
Macintyre, S., 255
Madood, T., 73
malaria, 6, 27, 133–4
Malinowski, B., 52n.6
Mans, Inge, 217
Mante, Norris Saakwa, 217n.9
maple syrup urine disease, 94
marriage patterns
  barriers, 55–6
  black population, Southern Africa,
    151
  ethnic groups, 76–7: migrants, 61;
    self-identification, 59
  and frequency of genetic disease, 7
  incest taboo, 29
  Northern Europe, 30
  Mediterranean, 31
  migrant ethnic minority groups, 28
  patrilineal endogamy, 34–6
  patrilineal exogamy, 32–3
  see also arranged marriages;
    consanguinity
Marshall, J. Richard, 202–3
Marteau, Theresa, 42, 202, 203
mastectomy, 187
maternal serum screening, 12
maternity services, 99
mating frequency, 55
matrilineal kinship patterns, 35–6,
  37
medicalisation, 252
Mediterranean kinship patterns, 31
Mendelian disorders, 5
Mendelian population, 55
Merton, Robert K., 252
Middle East, 31

migrants, 28, 61, 68–70
Modell, Bernadette, 40, 41, 52n.10,
  63, 84–5, 116
Molineaux, L., 136
Mozambique, 134
Muller, H.J., 249
multiculturalism, 45
multifactorial diseases, 5, 14
multiple sclerosis, 43
Murray, C.J., 250
muscular dystrophy, 125, 126, 169
Muslims
  antenatal screening, thalassaemia,
    112
  blasphemy laws, 72
  consanguineous marriage, 84, 91,
    114
  first-cousin marriage, 76
  and foetal sex selection, 36
  inherited disease and stigma, 41
  patrilineal kinship patterns, 34–6
  unconscious fantasy in genetic
    counselling, 160, 161
myths, 152–3

National Curriculum, 241
natural selection, 7
Navaho Indians, 150
neo-Nazism, 215
neural tube defects, 97, 198
newborn screening, 11, 15
Nigeria, sickle cell anaemia, 133,
  144
  political will, 136
  prevalence, 135
  Sickle Cell Clubs, 138–41
North Africa, 31
Northern European kinship patterns,
  29–30
Novaes, Simone, 226
Nuer, 67–8
Nuffield Report on Bioethics, 47–50,
  201, 233, 239
  eugenics, 240

Okeley, J., 69
oophorectmoy, 187
Orthodox Jews, 60
osteomalacia, 62